The Children of Time

The Children of Time

CAUSALITY, ENTROPY, BECOMING

Rémy Lestienne

Translated from the French by

E. C. NEHER

University of Illinois Press   URBANA AND CHICAGO

Les Fils du Temps © 1990 by Presses du CNRS.
English-language translation published in 1995 by the
University of Illinois Press under license.
Manufactured in the United States of America
1 2 3 4 5 C P 5 4 3 2 1

This book is printed on acid-free paper.

Library of Congress Cataloging-in-Publication Data

Lestienne, Rémy.
    [Fils du temps. English]
    The children of time: causality, entropy, becoming / Rémy
Lestienne ; translated from the French by E. C. Neher.
       p.   cm.
    Includes bibliographical references and index.
    ISBN 0-252-01959-8 (cloth : alk. paper). — ISBN 0-252-06427-5
(pbk. : alk. paper)
    1. Time. 2. Causation. 3. Becoming (Philosophy). 4. Entropy.
I. Title.
BD638.L38   1995
115—dc20   94-20269
                                                   CIP

Contents

Introduction

One might also raise the problem of whether time would exist or not
if no soul existed . . .

 —Aristotle, *Physics,* 223

Time is there, a haunting presence. We feel ourselves balanced precariously
on the crest of a wave that never breaks. An analysis of this impression
shows that it has several different components.

One of those components is related to our hunger for action. Upon
waking we might outline the things we would like to accomplish during
the day. In the evening, more often than not, we realize that our program
remains unfulfilled, and our expectations therefore engender a perpetual
dissatisfaction, exacerbating the sensation of the passage of time.

A second component concerns the testimony of our senses, which tell
us of a time that leaves its mark on the world around us, at least insofar as
our perceptions are to be trusted. This is the time of change, evoked by
visible movement or by audible melodies.

But do we even need the testimony of our senses? With eyes closed, the
melody silenced, let us allow our consciousnesses to float about in this
sensory wasteland: time still passes. This is the time of "pure duration,"
or "the form taken by the succession of our conscious states when our 'I'
lets itself live,"[1] which Bergson speaks of as the purest form of the intuition of the passage of time.

Once passed, time leaves the imprint of its teeth, persuading us that it
is neither an illusion nor a movement of the spirit. The world around us
not only moves but ages, as do we. We not only live *in* time: we *consume*
time, as a car burns fuel. Even this comparison is inexact, however, since
without gas a car remains a car. But without time all disappears as does an
image on the screen of a movie theater: if the film stops rolling there, it
leaves not just a still picture but a catastrophe; the film burns and the image is lost forever.

It may seem strange that an experience as basic as that of time and its passage should be the subject of a book delving once again into the historical discoveries of science while making use of the most recent findings of modern research. The appearances of daily life urge us to adopt a realistic position in the philosophical definition of this term—namely, to hold firmly that time certainly exists, independent of us and of our discourse. Nonetheless, it is in no way inappropriate to ask whether time is real or illusory. This passage of time, which we instinctively hold to be true, is in fact the subject of a centuries-old debate. And far from being in accord with common sense, two thousand years of philosophical thought and two hundred years of scientific thought point inevitably in another direction: that identified by Immanuel Kant, for whom neither time nor space belonged to the world in itself—the world as it exists, independent of our senses—but only to the world as conceived by the mind.

■ Our Access to Time Is Indirect

Who can deny that our knowledge of the external world is mediated by our perceptions and that organized perception—in which events are arranged according to the place and moment in which they appear—presupposes that our minds define a spatial and temporal framework? It is apparent, even when it comes to discoursing on the true qualities of this world immersed in the framework of time, that the departure point is the self, the symbolic interpretations we give to our perceptions. Because of this, we have some trouble in discerning and distinguishing what, in this case, is truly a part of nature itself from what belongs to us, what, in other words, could be no more than a reflection of the mechanisms of our central nervous system and mechanisms of perception. And naturally, we question ourselves about the universality of the flux that carries us onward and upward. The stars in the sky, the atoms, the objects that surround us, are they also *within* time?

Does time also play a special role in the phenomena of life? Does it play an active role in the development of the individual, in maintaining the day-to-day vital functions, in the phenomenon of consciousness, of intellect, explained—as much as it can be—in terms of nervous mechanisms?

Here we find ourselves squarely in an area where philosophical questioning preceded a strictly scientific contribution. Today we know to what a large extent consciousness can act as a prism distorting our raw sensations, filtering and adapting them to our hidden needs. We know that the two hemispheres of our brains receive different sensory impressions, which are integrated into one unique and coherent perception only at a relatively high level of cortical activity. The American neuropsychologist Roger Sperry has shown that the discordant sensations received respectively by the right and left brains can be both unconsciously perceived and recognized but also rejected, refused at the level of the conscious when they do not coincide with the dominant impressions of the subject.[2] We have then, in our minds, a structure that sorts through sensory impressions, decides on their pertinence, and transforms them, perhaps without our knowledge. Being conscious is not only being conscious of the world around us or having an internal representation of it. It is, above and beyond all, the moment-to-moment organization of these sensations according to the need for unity and coherence that permits us to be aware of ourselves, in our individuality, at the crossroads of these perceptive modalities. Being conscious means always insisting on being *one* in *one* world, even at the cost of uncontrollably distorting or more or less severely culling the raw information furnished by our senses. Haven't we the right, therefore, to doubt the evidence of our senses concerning the passage of time?

■ A Preliminary Definition of Time

All kinds of hidden dangers threaten a discourse on time. Forgetting how dependent our comprehension of time is on our cognitive processes, on our mechanisms of perception and memory, is one of those dangers. We can, however, formulate a preliminary hypothesis that time does indeed exist independently of us, that it is an essential dimension of reality. A second obstacle then would be the desire to identify a priori that independently existing time with causality or entropy, the two concepts that science has successively elaborated to apprehend (and interrogate) two important facets of the fundamental notion of time. Do these concepts apply to nature itself or rather to our observations, that is to say, to nature already put to the question by our instruments of measure, interpreted by our senses, and classified according to abstract categories? Moreover, each of these concepts may describe only one or another aspect of time, not its totality. For this reason the definition of time as a primary notion must be the least distorted and the least constraining possible, as long as, after examination, it survives a confrontation with current scientific fact.

The following is one possible definition: *time is a degree of freedom by use of which objects retain their identity while displacing themselves in a manner called "future."* This definition might be judged to be either excessively vague or already too daring. It does, in fact, presuppose the acceptance of some sort of physical reality, namely, the existence of objects other than ourselves. All scientists would doubtless agree on this point. How else to escape from solipsism, from that doctrine that states that nothing is certain but our own sensations and that nothing exists outside of those sensations? Solipsism seems incapable of explaining why, when faced with an event experienced in common with other witnesses, our sensations seem to be so like those of our fellow witnesses.

The above definition of time equally implies the idea of degrees of freedom. In physical sciences, this notion refers to the number of a system's discrete states whose magnitudes may vary. For example, a particle trapped in a container can occupy several positions, and its velocity can be oriented in different directions and assume several different values. But it can also exist at several successive instants in relationship to a clock of reference. Of course, the particle doesn't "choose" the indications of the clock, and freedom thus understood has nothing to do with a free choice on its part. In the case of time, what is important is that the degree of freedom cannot be manifested in more than one direction. In other words, time has a direction. Outside the concrete possibility of realizing this degree of freedom in the indicated direction, time does not exist. We must guard ourselves against succumbing to the enchantment cast by those who raise the possibility of an inversion of time in the fantastic regions of the universe near the infinitely big or the infinitely small. Time is defined as a function of its arrow. If the potential liberty of objects to go toward the future did not exist, or if it were inverted, we could no longer speak of time. Before the "Big Bang" or after the "Big Crunch" there are not yet or will no longer be "objects" that can retain their identity while moving through time. How could one construct a discourse on the absence of time that would, so to speak, be still immersed within the framework of time? In some interactions, certain subatomic particles seem to obey final laws, so much so that some physicists have described them as "moving backward in time." This description, which evokes a time "in reverse," seems self-contradictory. In describing the phenomena in question, it would be much better to use a category other than that of time and to call on an appropriate adaptation of the definition of causality.

■ Time as a Dialectical Concept

Time as a concept is complex in two senses of that term. In the first sense, something that invokes contradictory aspects is complex; in the second, that which is complicated, which brings a large number of elements into mutual interaction, is complex.

Consider first the contradictory aspects of time. When we speak of time, our speech immediately invokes, sequentially or simultaneously, the two concepts of *permanence* and *change.*

In all human languages permanence lies at the root of the invention of nouns. In naming objects, including our bodies, we abstract from many instantaneous sensations to form concepts of that which those sensations share. The world, however, is not static: the ball rolls, the river runs, the wave breaks, the stars continue in their courses, atoms disintegrate, and bodies age. The intuition of change must, therefore, be added to that of permanence. In fact, it is not at all certain that, in our first experiences of the outside world, change is not more important than permanence. We know, thanks to the discoveries of Hubel and Wiesel, that some cells of the visual cortex are much more sensitive to variations in stimulation, to objects in movement, than to static objects.[3]

The development of science has given birth to two temporal paradigms that parallel this "dialectic" of permanence and change: *causality* and *entropy*. I will introduce them here in order to underline this parallelism.

When Galileo discovered the fundamental importance of time, he gave the decisive impulse to the scientific adventure of deciphering nature's language. The elaboration of classical science since Galileo, thanks to Newton, Maupertuis, Lagrange, and even Einstein, has revealed one of the essential reasons for the success of introducing time into physical theory. Time and space reveal, together, a formidably powerful paradigm for the organization of the material world: the paradigm of causality, which is closely tied to the category of permanence. Indeed, causality derives from the conservation of energy, an immutable quantity in any closed system. It also indicates the inevitable course of things, the absence of surprise. It proclaims, finally, the absolute equivalence of causes and effects, of the future and the present.

Contrary to the peremptory affirmations of some of its most fervent promoters, however, the paradigm of causality and the concepts of space-time associated with it do not exhaust the original concept of time. In fact, they neglect its essential aspect: becoming, aging, the fact that things follow constantly and with single-minded purpose the gradient of increasing entropy defined by Rudolf Clausius.

The recognition of this aspect of time-becoming as a genuine ingredient in scientific explanations dates to the end of the last century; today it is seeing a rebirth of interest. It is at the core of the thermodynamics of irreversible processes and is beginning to be applied also to life sciences. In this context, time appears stamped with the seal of complexity. When we use our instruments to investigate the behavior of the most simple objects of nature, the elementary particles, atoms, and simple isolated systems, time's arrow does not appear. It shows itself, on the contrary, in macroscopic systems considered from a practical point of view by thermodynamics. Currently, physical theory remains incapable of explaining temporal irreversibility, except at the extreme limits of systems of infinite size that are beyond our reach.

■ Is Man a Time-Perceiving Machine?

Philosophy and psychology teach us that the perception of time is a pre-condition of our elaborated worldview. Is the sense of time specific to humankind or especially sharp in us? Is there an essential difference on this point between humans and animals? If time exists objectively, and *Homo sapiens* possesses a special acuity for this dimension of nature, discussing time takes on a double virtue: that of speaking of man when speaking of the universe.

Experimental science has underlined the grand continuity that reigns in the living world: chemical substances turn out to be identical or very close cousins, cellular organization and the structure of organs resemble each other, the general design of organisms is often comparable. The findings of paleontology reveal that the human adventure began some millions of years ago, when certain anatomical modifications (the position of the skull on the spinal column and development of the cerebral cortex and the cerebellum), giving rise to new attitudes and the acquisition of new faculties, that of speech in particular, marked the rapidly growing distance between the genus *Homo* and the other primates. On the scale of behavior and cerebral capacity, however, when experimental psychologists and ethologists try to evaluate the differences engendered by these minor anatomical modifications, they are struck by how such small changes resulted in such important consequences. For example, in tests of their capacity for anticipation, animals exhibit very mediocre performance once the time periods involved surpass a few minutes. It cannot be ruled out that humans have crossed a threshold and that their aptitudes result from a qualitative leap. But the nature of this threshold still remains obscure.

Even in the demand for unity that characterizes human consciousness, an abstraction made from the psychological intuition of the flow of time, the problem of duration remains. The viewing of a scene, the hearing of a sound, the feel of an object all require a time lapse in order that sensory impressions be processed and give rise to a conscious perception. In the raw state, psychological perception is, therefore, constructed of a choppy series of perceptions: psychological time is fragmented. At the level of a higher consciousness, however, these different facts must be reintegrated to allow a continuous and harmonious representation of the self and of the world. It would be impossible, otherwise, to reestablish in our thinking the continuity of time, which is necessary for the existence of motion and the interpretation of the world according to physical paradigms such as causality. This function of temporal integration is shown to be even more important when longer periods are at issue: the "I" is founded on the intimate conviction that we retain our identity not only in the instant that is to come but also later, tomorrow, or in several years. This heightened consciousness of even distant time constitutes both our strength and our drama. Alone among the animals, man can ask himself what he will do tomorrow. Here, without doubt, is a quick formula to define the specificity of the human mind. In this sense, the ability to think about time is an eminently human faculty.

This is not to say that time *itself* is a pure invention of the human mind.

Part 1

Where Do We Get Our Ideas about Time?

1

The Myth of the Eternal Return and the Concept of Progress

Paleontology cannot reveal to us the exact thoughts and fears of the first humans. Certainly they were already curious about time, since the development of thought employing a temporal dimension is one of the characteristics of the emergence of humanity. "The eyes of both of them were opened," says the Bible, "and they realized that they were naked." Looking at this allegorical text in terms of temporality, we can envision early humans stricken with vertigo by the signs of the grand tide constantly changing their view of the surrounding world. Incapable of determining whether it is the world observed by the immutable human mind that is incessantly metamorphosing or the changing mind that follows an abstract trajectory across an immutable stage, they try to hold onto a fixed point, to some certainty, to some solid frame of reference that will allow them to believe that their existence is not an illusion. Imagine the relief these early humans must have felt on discovering, at the core of their variable daily experiences, the existence of certain events that continuously repeated themselves, such as the beat of their pulses, the alternation of day and night, the return of the seasons. These repetitive experiences, marked by a cycle or by a rhythm, must have been among the first experiences to reassure our earliest ancestors, allowing them to set their minds at ease.

The evidence is that these cycles or rhythms conferred a new quality to the events they linked, one that transcends irreversibility and confers a character of stability and permanence. These cycles are the building blocks of identity emerging from change. They open the door to comparisons, as well as to a measurable and communicable definition of time.

During the rise of classical culture, cycles and rhythms were called upon to play an even more important role. Through them, duration, a quality linked to fleeting phenomena, became a myth or even a person. Time itself came to be revered as a god.

■ The Social Role of Cycles and Rites

Cycles and rhythms were the instruments, the mediators, of the intercommunication and socialization of time. This first stage was quickly superseded by the use of *rite,* the social expression of rhythm and cycle. Rite profoundly marked the first forms of society's temporal organization. From ancient times to the Middle Ages, the need to regulate the times of the liturgy and the hours of prayer constituted a powerful motivation for observing the stars and computing astronomical tables.

Liturgical time, however, entailed the major inconvenience of unequal hours. Little by little the needs of civil organization and economic production, as well as the contributions of technical progress (construction of increasingly regular and precise clocks), allowed first the invention in the high Middle Ages, and then the imposition, of a civil time with a day divided into twenty-four equal hours, followed sometime later by the division of these equal hours into smaller and smaller portions: minutes, seconds, and so on.

Just as religious imperatives prompted a socialization of time, philosophy took a major leap forward once this concept expanded to fit the needs of society. Escaping from the circle of time, breaking the hegemony of the cycle, it substituted a time that was open, progressive, and linear. In this new form, time could be reinvested with the qualities of irreversibility and incessant becoming, qualities the ancient world had stripped from it. The notion of a linear time easily carried the day. It still forms the basis of our cultural universe, close to familiar notions of space and distance and in tune with the intimate intuition of the passage of time.

This transformation took place in large part under the Athenian republic. In ancient Greece both the cultural world and the practice of civil order suggested a circular model of nature and a cyclic model of time. In his book *Myth and Thought among the Greeks,* Jean-Pierre Vernant[1] exposes the astonishing power of this myth of circularity, which held sway over concepts of both space and time. In terms of space, it imposed the circular organization of the city around the marketplace (the *agora*). In terms of time, the eternal round of day and night and the regular succession of the seasons imposed the adoption of a perpetual calendar.

This idea of cyclicality was already present in the poems of Hesiod in the eighth century before Christ, which illustrate the cyclic concept of time intimately linked to the practice of agriculture. It became a central theme in the philosophical tradition through the writings of Plato, teacher of Aristotle and father of occidental philosophy. In Plato's works the myth of circularity finds its source in the apparent movement of the celestial sphere. This movement, eternally present but totally inaccessible, was witness to the true essence of the world, which consisted not of perishable objects but of eternal Ideas (or Forms). The inexorable movement of the celestial vault was not eternity itself, of course, but the signature of eternity. Plato writes in the *Timaeus:* "Now the nature of that Living Being was eternal, and this character it was impossible to confer in full completeness on the generated thing. But he took thought to make, as it were, a moving likeness of eternity; and, at the same time that he ordered the Heaven, he made, of eternity that abides in unity, an everlasting likeness moving according to number—that to which we have given the name Time."[2]

The umbilical cord that links time considered as a measurable entity to its opposite, to the denial of irreversibility that constitutes the myth of eternal return, cannot be more strongly underlined. Deprived of the knowledge of perfect things such as eternity, we cannot apprehend *becoming* except by way of the cycles, by counting them, and we cannot conceive of eternity other than through change. Learning from these tangible signs, which are only shadows of real things, we must, to arrive at a knowledge of truth, disregard them and hold tight to the world of certain and eternal Ideas, of which the world of tangible things is only a reflection.

Of course, the modern world has rejected this Platonic idealism, which discourages any kind of scientific approach. But to forget this paradoxical origin of measurable time to which Plato had already drawn attention would be detrimental to understanding the modern concepts attached to time. To grasp the flow of time we count off and enumerate the cycles, each of which is like a challenge to time. The standard by which we measure time is that of cycles devoid of *becoming,* in other words, bits and pieces of anti-time.

■ The Time and Motion of Aristotle

Maybe it was this paradox that led Aristotle to redefine time by freeing it from such impure origins. Plato's celestial canopy is neither time nor the reflection of time but the reflection of nontime. The stars do not belong to this world; they are not subject to the stringent law of change. They belong to the world of perfection and eternity, and their substance does not suffer corruption. Their uniformly circular movement, which engenders cycles, belongs not to time but to eternity. Thanks to their eternal return, we can contemplate the stars as jewels set in the crown of eternity.

It is hard to imagine the effort of abstraction Aristotle must have made to dissociate time from circularity and to seek, with the aid of *motion,* a new foundation for the concept of time. In the Athenian republic, everything led back to the circle and the cycle, perfect forms of space and time. Aristotle had to wrench himself free of these familiar models to affirm that the world of cities and people, up to and including the Moon, obeyed other norms, other forces, to which no one had as yet drawn attention.

Much later, Galileo and those of his successors who articulated the law of inertia in its final form performed a comparable effort of abstraction when they attempted to grasp the essence of inertia by abstracting terrestrial gravitation from everyday experience. In truth, Galileo himself was only partially successful. According to him, a ball rolling on a horizontal plane of infinite extension retains its movement indefinitely. The infinite horizontal plane, however, corresponded in his mind to the spherical envelope of the Earth at the altitude under consideration. It is only the complete elimination of gravity that allows an exact enunciation of the law of inertia. And this abstraction, begun by Galileo, was completed only by his successors Gassendi and Descartes, as has been shown by Alexander Koyré.[3] Revolutionary moments like these are rare in the history of ideas, moments where it is necessary to reject ancient ties and to dare to propose new sources for familiar concepts in order to escape a contradiction.

But what then of Aristotle's approach to a time freed from circularity? The point of departure of his concept rests on the experience of motion. Motion, in and of itself, cannot be measured. Beyond the experience of motion, however, a measurable concept of time can be constructed. Thanks to motion, in fact, we can measure time in the same manner as we measure space, by distance traveled. This is the entire program of classical mechanics in a nutshell, fully realized only in the works of Galileo, Barrow, and Newton. But its seed is found in the reflections of Aristotle, freed from the mythical thinking of Plato and already approaching the spirit of modern physics.

First Aristotle affirmed that time and motion are related but distinct notions. Time is not motion "but something of motion." Motion is found in the moving object, whereas time is everywhere and identical in each thing. Moreover, motion can be faster or slower, whereas time always passes at a uniform rate.

In other respects, however, time and motion share certain characteristics: both are continuous and successive. The continuity of motion flows from that of space, from its infinite divisibility. The continuity of motion leads in its turn to that of time, because one cannot conceive of motion without the continuity of time; this is shown by the paradoxes of Zeno of Elea. There existed, therefore, according to Aristotle, a relationship between the continuity of space, of movement, and of time.

Moreover, time is successive, that is, it repeats itself at each instant as it flows, as does motion. Aristotle emphasized the distinction between the before and the after. Between two moments, between two snapshots of motion ordered according to anterior and posterior, something he called *time* had passed. Hence his definition: "time is a number of motion with respect to the prior and the posterior. . . ."[4] It becomes a question then of a mathematical and objective time that orders two successive moments and that allows, step by step, the labeling of a succession of moments according to the succession of increasing integers.

It must be emphasized that Aristotle made no allowances for technical concerns in this approach to a conceptualization of time. He did not consider how to measure time concretely. As a theoretician, he discovered the means to define time as enumerable quantity, but he avoided speaking of duration as being technically measurable, since that would entail falling back on the use of cycles. He did attempt it in a later stage, however, after having affirmed that, contrary to appearances, once a cycle has run to its end and all things have returned to the same place, time has indeed changed.

Thus, for Aristotle, time was a concept made necessary by the reality of motion. It was the number—that is to say, the reference or the axis of coordinates—to which motion must be related, similar to the way in which space constitutes the reference, or the system of axes of coordinates, to which expanse is related. In this, the ideas of Aristotle are already remarkably close to those of physicists such as Galileo, Newton, and Descartes. They even evoke, in some respects, the positions of Kant on a priori categories. For all this, the modernity of Aristotelian time must not be exaggerated. It broke apart the inexorable cycles and allowed escape from the curse of the Eternal Return, but it would take another nineteen centuries for classical physics to conceive of an "absolute, true and mathematical time" (in Newton's words), axes of coordinates similar to those of space, open to the infinite.

The prodigious feat of abstraction Aristotle accomplished carried with it its own limits. His concept of time *is* mathematical, but it is also terribly abstract. The instant, which he developed extensively, is understood as a mathematical limit between the before and the after, between the past and the future. He did not specifically examine a means of attaining this limit, almost certainly because of the imprecision of the contemporary instruments (water clocks, hourglasses, and sundials). Above all, Aristotle did not examine the relationship between this vanishingly small slice of time—the instant—and motion. The idea he formed of velocity is, all things considered, extremely poor. In both his teachings and his writings he distinguished rapid from slow motion. He understood the concept of uniform velocity but not that of instantaneous velocity.

By breaking open the circle of time, Aristotle did allow the birth of the modern notion of time: a human, historical time related to the human experience of motion and of the endless flight of time as it is lived. He himself addresses the question: "It is also worth inquiring how time is related to the soul," he declared. "So if nothing can do the numbering except a soul or the intellect of a soul, no time can exist without the existence of a soul, unless it be that which when existing, time exists, that is, if a motion can exist without a soul."[5] These labored ruminations show that Aristotle was aware of the close relationship between physical time and time as it is lived. Moreover, in the world of change, things, as well as humans, obey a destiny. In light of this background, there is nothing strange in Aristotle's having made such a strong distinction between the sublunar and the supralunar world: to the latter, the world of static or cyclic perfection, he opposed the former, the world of change or becoming. True time is that of the changeable world, which is found under the Moon, the unstable world in which each thing tries, incessantly, to return through motion to the place that destiny has assigned it.[6]

Aristotelian time therefore already possessed the principal characteristics we today are tempted to attribute to this concept of time. Drawn from the time of consciousness for its enumerability (therefore its "arrow"), it belongs to physics under the double heading of motion and becoming. But did Aristotle already represent it to himself, as would Newton, as an analog of space, an axis of coordinates, a straight line running from minus infinity to plus infinity? This would, of course, be reading too much into his feat—too much for Aristotle to accept and too much even for his successors, the great Arab physicists who were the concept's interpreters and mediators and among whom are to be found Avicenna (whose conceptions of time were closer to those of Plato) and al-Ghazali.

The decisive factor in this battle between cycle and indefinite becoming might have been, as suggested by G. J. Whitrow, the appearance of Christianity and the historical notions of redemption and personal salvation, both irreconcilable with the myth of the Eternal Return. Did not Saint Augustine, Neoplatonist that he was, write in *The City of God:* "So that, by following the straight path of sound doctrine, we escape, I know not what circuitous paths, discovered by deceiving and deceived sages."[7] Hervé Barreau retraces in detail the history of linear time's slow maturation.[8] It is a maturation that passes through the writings of the Arab and medieval philosophers who transmitted Aristotle's doctrine to the Renaissance; on through the writings of Averroës, who, from the lofty position of his chair at Cordoba in the middle of the twelfth century, seems still to have refused to accept the representation of time as a straight line; and then to the writings of Thomas Aquinas, who finally reconciled Aristotelianism and Christianity a century later. The notion of a rectilinear and indefinitely extended time, similar to space, did not appear until 1690, in the writings of John

Locke. "But yet there is this manifest difference between them, That the *Ideas* of Length, which we have of *Expansion, are turned every way,* and so make Figure, and Breadth, and Thickness; but *Duration is but as it were the length of one straight Line,* extended *in infinitum.*"⁹ By this time, of course, Galileo and Newton had already presented their discoveries to the world. Nonetheless, in the twelfth century, long before Einstein, the Jewish Arab philosopher Maimonides, following a dazzling and premonitory intuition, was already teaching at Cordoba that time is only a manifestation of matter, nonexistent before it!

2

The Invention of the Concept of Instantaneous Velocity

Despite the break in the circle of time, medieval time remained too closely linked to the concept of cycles to give birth to modern mechanics. The gap existing between the instant—the abstract limit between before and after—and the uniformity of cyclical movement was not easily bridged. Nothing illustrates this difficulty better than to study retrospectively the birth of the concept of velocity.

For twentieth-century individuals, instantaneous velocity is a familiar concept, as intimate as the speedometer reading of a car. This familiarity is due in large part to the modern lifestyle, to the role played in our occupations by all the artificial means of locomotion and to the incessant battle we fight against lost time. Some psychologists, thrown off the track perhaps by these peculiar circumstances, thought they saw in velocity a concept even more elemental than that of time, to such an extent that they believed the one derived from the other in the course of a child's growth: "we cannot see or perceive time as such since, unlike space or velocity, it does not impinge upon our senses."[1] This conclusion, taken from a study on the psychological development of children in contemporary civilization, is not confirmed by studying the evolution of the concept of velocity; the history of science shows rather that the acquisition of this concept in its modern sense was both difficult and laborious.

Again, velocity is an abstract parameter associated with a moving object that lets us describe the object's motion at a given instant in its trajectory. It is defined in physics as the limit, for intervals of time approaching 0, of the ratio between distance traveled and the time taken to travel that distance. In order to grasp the concept fully, the possibility of dividing the distance covered by a corresponding duration must be allowed. But since classical physics did not view duration and distance as being measurable by common standards, it therefore became necessary to introduce standards other than those of expanse and time. It was also necessary progressively to reduce the measured or imagined intervals to infinitesimal proportions. To do that, it was necessary to understand and discuss the concept of the instant as related to duration.

■ The Nature of the Instant

Any understanding of duration demands consideration of a segment of lived time within which final and initial conditions can be compared—hence the difficulty in conceiving of an interval of time approaching 0. Bergson, however, clearly expressed the feature that distinguishes conscious time from space: the present contains the past, whereas a point on a line contains nothing of the point or points that precede it. Some philosophers and epistemologists regularly go so far as to propose introducing an atomicity of time. Intervals of duration and mathematical instants thus seem of an irremediably different, if not opposite, nature. From this point of view, the epistemological program of deducing the latter from the former by letting t tend toward zero might appear to be an inappropriate enterprise. It was so, certainly, in the eyes of the ancients, and even more so since their instruments for measuring time—sundials, water clocks, and hourglasses— were intended not to measure brief intervals but rather to mark the peaceful passage of the hours.

The difficulty in conceiving of instantaneous velocity and the absence of appropriate chronometric techniques in large part caused physical theory to stagnate between antiquity and the Renaissance. To surmount this difficulty, it was first necessary to take recourse in an artifice: to replace, whenever possible, real movement at variable speeds with an equivalent ideal movement at a constant speed. The history of physics before and until the time of Galileo shows this clearly. It was by means of this artifice that the physicists of the Middle Ages and Renaissance, after centuries of Aristotelian tradition, were able to come progressively closer to the study of motion at variable speeds and to the study of falling bodies.

■ The Analysis of Physical Motion

In his discourse on physics, Aristotle discussed the conditions in which a body could find itself in a state of motion. He distinguished natural motion (due to gravitation) from violent motion (produced or altered by a force imposed on the body) and applied himself to refuting the arguments proposed by the disciples of Democritus for whom motion presupposed the existence of a vacuum. Aristotle distinguished rapid from slow motion, but he neither dissected motion nor analyzed it. He considered it as a whole, hence the limited character of his concepts of velocity.

Looking at falling bodies, Aristotle acknowledged that the velocity of an object in free-fall depends only on its weight and the resistance of the medium in which it travels. It would be, therefore, a constant of the body's fall for any given medium. This constant was tied to the conditions of the experiment: it belonged at the same time to the body and to its environment. Specifically, for a given medium traversed, the velocity associated with a body's natural motion, which measures the body's propensity to return to the place assigned it by the natural harmony of the world, depends only on the body considered. It constitutes a characteristic of the body (like mass), not a kinematic variable.[2]

Velocity characterized as a constant of motion certainly influenced the Greek astronomers, and in particular Ptolemy when he formulated the theory of epicycles: the nonuniform appearance of planetary motion must result from the combination of several circular motions, each of which was described at a constant velocity. Thus, the law of uniformity could, in some sense, be safeguarded. It sufficed to consider several uniform motions: that of the planet in its epicycle, that of the sphere at which the center of the epicycle is located, and so on.

The first allusion to velocity as a kinematic variable probably must be attributed to William of Occam, who around 1320 asked himself what distance a body in motion would travel in a given time if, at a specific moment, the motion were allowed to continue without alteration. Whatever the case, a great distance remained between this first intimation and the correct expression of velocity as defined by Newton.

It can be instructive to follow the different stages of this discovery, to witness an increasingly exacting breakdown of motion. Around 1350, Albert of Saxony distinguished three phases in describing the motion of a horizontally projected stone (fig. 1):

1. At first the stone effects a "violent motion" (in Aristotle's sense) under the influence of the primitive horizontal impetus. During this period the trajectory is horizontal since, according to Aristotle, two motions cannot coincide in the moving body.

2. However, air resistance gradually consumes the "impetus" initially imparted to the stone. At this time a short transitional period comes into play during which the violent horizontal motion gives way to the natural vertical motion, in which gravity attracts the stone to its natural resting place. Although he insisted on its brevity, Albert of Saxony did not discuss this transitional period, which was mysterious because it implied a superposition of the two motions.

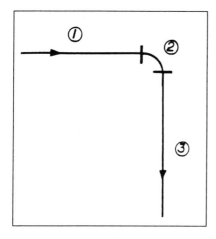

Figure 1. The three phases in the motion of a stone thrown horizontally, according to Albert of Saxony. In (1) the violent motion imposed by the hand imparts a purely horizontal motion to the stone. In (2) the initial impetus is exhausted and gravity diverts the stone downward. In (3) gravity acts alone; the stone pursues a rectilinear downward motion.

3. Finally a third phase is seen once gravity has recovered its preeminence. The stone then follows the motion characteristic of weighted bodies, which Albert of Saxony did not analyze further.

At about this same time, Oresme undertook the study of nonuniform motion. He used the rule of the composition of velocities (the rule of the parallelogram, according to which diverse motions can coexist in the same body, the corresponding velocities then being added vectorially) already introduced during the preceding century by Jordanus. Consequently, he clearly abandoned Aristotle's prescription as to the unity of motion. He very explicitly used the rule of the composition of velocities to study the motion of bodies subjected at the same time to a natural motion and to a "violent motion." Notably, he refuted the objection advanced at the time denying that the Earth turned. An arrow shot vertically into the air, said the partisans of the Earth's immobility, falls back at the archer's foot. Now, if the Earth were turning while the arrow flew, the Earth would continue to turn and the ground to move, so that the arrow would not fall at the feet of the archer. Oresme responded that the loosed arrow, like the archer, participates in the motion of the Earth and that this motion of rotation must be constantly added to the vertical motion imparted to the arrow by the archer and the upward or downward effects of gravitation. The arrow will come back to the exact point from which it was fired because the horizontal component of its motion is identical to the Earth's horizontal motion.

It was still in the fourteenth century that the modern expression of velocity as the ratio between distance traveled and time elapsed first appeared in the writings of scholastic mathematicians. Oresme studied, for the first time, uniformly accelerated motion, in which the moving body's velocity increases or decreases regularly over time.

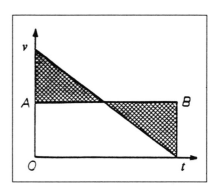

Figure 2. Oresme's proof that distance covered in a uniformly accelerated motion (here decelerated) is equal to that covered in a uniform motion at velocity $v/2$ (the average between the two extreme values of velocity during the motion). Oresme's intuition is confirmed by modern integral calculus as a consequence of the equality of the areas of *OABt* and *Ovt*.

He represented these motions by a straight line on a diagram in which velocity and time are given respectively on the y-axis and the x-axis (fig. 2). Oresme showed that the distance traveled by the moving body between the instant 0 (velocity v) and the instant t (velocity 0), given by the surface of the triangle *Ovt*, was the same as that traveled by the moving body given a uniform motion at velocity $v/2$ (surface *OABt*). Oresme thus fused the study of uniformly accelerated motion with that of uniform motion. This technique opened the door to the study of more general variable motion, which could always be divided into segments of time short enough that, during each segment, the moving body could be considered to be animated by a uniform motion.

Galileo would make systematic use of this idea when he began to study free-fall.[3]

■ Galileo and Falling Bodies

By using Oresme's argument, Galileo showed, in a letter addressed to Paolo Sarpi in 1604, that a body subjected to a uniformly accelerated motion travels distances that are, among them, similar to odd numbers—1, 3, 5, and so on—when that motion is divided into segments of equal time. In other words, the distance traveled since the instant 0 increases as the square of the time elapsed (fig. 3).

It must be noted that, in this description, Galileo subdivided accelerated motion into segments of time such that, during each segment, the significant quantity is the average velocity of the moving body.

It is at precisely this stage, when motion was being divided into smaller and smaller segments of time (foreshadowing the Newtonian idea of velocity as the limit of the ratio between distance traveled and elapsed time), that the parallelism between the evolution of ideas about time and the evolution of chronometric techniques becomes evident. In fact, conceiv-

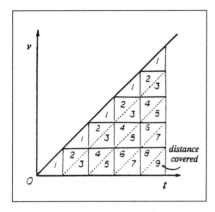

Figure 3: Sketch of the drawing used by Galileo in his letter to Paolo Sarpi on falling bodies. Each elementary triangle represents one unit of distance traveled by the moving body. One can read directly from this graph that distance traveled in equal amounts of time follows the sequence of odd numbers.

ing of smaller and smaller segments of time comes down to imagining, if not to measuring, shorter and shorter intervals of time. It is not certain that Galileo himself actually conducted his famous experiments on the inclined plane. He described them, however, in great detail, as if he were trying to convince the reader of the practical possibility of the dissection to which he submitted motion.

Galileo used a wooden ball, an inclined plane (a plank of wood grooved in the middle so that the released ball would follow the line of greatest inclination), and, as a chronometer, a water clock (a vase full of water pierced at the bottom and equipped with a tap). The experiment consisted of releasing the ball and simultaneously opening the tap, then closing it at the instant at which the ball had covered the entire incline. Later the experiment was conducted using only one fourth the length. The weight of the escaped water measured the time elapsed. Galileo observed that in this second experiment, half as much water ran out, and therefore half as much time had passed. As a consequence, the distance traveled varied as the square of the time elapsed. The technology did not allow him to do any better, and he did not seek to define a more exacting experiment comporting a finer subdivision of motion.

■ Instantaneous Velocity and Advances in Clock Making

In the fifteenth century technical progress in the field of clock making would be decisive in advancing the conceptualization of velocity. Bernard Walther used a mechanical clock equipped with a fifty-six-toothed wheel, which allowed him to work with intervals of time on the order of a minute (conceptually, minutes and seconds had been around since the fourteenth century).

But it was only after the application of the pendulum to clock making, thanks to an idea of Galileo's (Huygens, 1657), that true timepieces became available. In 1658 Huygens produced a pendulum clock that had three hands: next to the hour and minute hands appeared a new hand to measure seconds. Over the following decades the generalization of measurements of time as small as a second followed on the development of watches with springs and balance wheels. It was at precisely this time that Newton eventually discovered the kinematic concept of instantaneous velocity.

Galileo's Remarkable Error

The philosopher-savants of the Middle Ages had, as we have just seen, great difficulty in conceptualizing an instant. The consideration of ever briefer and briefer time periods, down to the instant, demanded an effort of abstraction all the more arduous because these thinkers could not depend on technology to help them make this leap in understanding: the measurement of short temporal intervals was beyond the capability of their timepieces.

The primitiveness of their era's clocks was doubtless partly responsible for the slow pace at which the founders of modern science became aware of the fundamental role time had come to play in physical theory. In that regard, no example is more enlightening than the lengthy procedure Galileo followed in uncovering the law of falling bodies. In effect, he had to abandon purely geometrical descriptions then in vogue and assign to time a central role in the description of motion. Galileo was not alone in his trials and initial errors on this subject. Leonardo da Vinci and Descartes, to cite only those mentioned by Alexander Koyré,[1] made similar errors in that same period. The repetition of those errors illustrates quite well the difficulty of graduating from a theory of nature founded on space to a description founded on time and the importance of this turning point in the history of scientific thought.

■ Galileo, Critic of Aristotle and Disciple of Archimedes

Galileo, who lectured in physics at Padua in 1595, spoke of "natural" motion, which, according to Aristotle, drew everything to its own quasi-organic location and in particular drew all weighted bodies to the Earth's core, as close to and as deep as their impenetrability would allow. At that point he was already criticizing Aristotle, who either did not search for or was unable to discover a quantitative description of motion and who satisfied himself with a final cause, as if objects obeyed a destiny, as do humans. His proud spirit led him perhaps even then to confide to some of his disciples his doubts on the pertinence of Ptolemy's cosmological system, which the school made him teach. He attracted his listeners with the freshness of his discourse, the fire he brought to his speeches on the possibility and even the necessity for a quantitative description, and his faith in the possibility of interpreting natural phenomena in terms of mathematical symbols and in reasoning about nature with the help of geometric figures. Cannot one hear in this speech an intimation of Einstein's exuberant exclamation, marveling at the mathematical intelligibility of nature?

Galileo had indeed read not only Aristotle and Plato but also Archimedes, who would be his principal inspiration. Perhaps he held in his hands Archimedes' treatises in their original Greek version, recovered at Constantinople and preserved until the sixteenth century in Italy by the family of the Italian humanist Lorenzo Valla. Imagine the fascination such a reading would exercise over him, a reading through which he discovered that geometry is the art, the means, and the space in which reason finds its surest guides, far removed from the verbose and obscure discourses of scholastic philosophers. At its core, nature obeys the principles of rational deducibility as they are to be found in geometry.

Galileo uncovered a veritable philosophy of scientific method in Archimedes' teachings: in physical theory one must not seek to account for the appearances of phenomena in too much detail. These appearances can be deceptive. Subjacent realities obeying mathematical rules can hide under a veil of contingencies, of chance, of perturbations. The nature to be unveiled is that ideal nature that the physicist unifies using the perfect science of geometry. One must therefore search for principles that form the essence of things and are susceptible to mathematical interpretation. An understanding of these principles will explain the phenomena, and obscure details will be illuminated.

Galileo's epistemology is clearly enunciated in the polemical work *The Money Assayer*, which he published in response to the attacks of his enemies: "Philosophy is written in that great book which ever lies before our eyes—I mean the universe—but we cannot understand it if we do not first learn the language and grasp the symbols in which it is written. This book is written in the mathematical language, and the symbols are triangles, circles and other geometrical figures, without whose help it is humanly impossible to comprehend a single word of it, and without which one wanders in vain through a dark labyrinth."[2]

When he first began to study falling bodies, Galileo immediately took this approach, using Archimedes' reasoning as a model. He was seeking a principle. From this principle would flow, through analogical transposition, a geometrical figure that would enable him to discover the law of motion. Such was the basis for the method he made his own.

In his study of floating bodies, Archimedes had introduced the principle to which his name has remained attached: the push exercised on a floating body is equal and opposite to the weight of the displaced water. Why should it not be thus also with air? Galileo therefore chose to apply Archimedes' hydrostatic principle to falling bodies. He developed this idea in his academic work *De Motu,* written some years earlier at Pisa. In *De Motu* he demolished Aristotle's argument that distinguished heavy objects, attracted downward, from light objects, attracted upward. All this, Galileo explained, is a matter only of a difference in specific weight: the object heavier than air falls; the object lighter than air rises. The velocity with which heavy objects fall is therefore a function of their relative specific weights. If all the objects that surround us fell in a void, their fall would have only one cause (gravitation), and they would therefore all fall at a uniform velocity, since the cause of their motion is constant (a constant cause, as Aristotle taught, can lead only to constant effects). But we know from experience that objects do not fall at a uniform velocity: on the contrary, they fall at a velocity that progressively increases, at least at the beginning of their fall. This follows from the principle proposed. At the beginning of their movement, objects plunged into air have "stored up" a lightness equal to their weight (equal, since they were at rest). This "virtue" stored up before their fall—lightness—opposes and slows their motion in its initial phase, but it is progressively used up, so that the overall result is an acceleration of motion.

Alas, the proposed principle did not lead to a clear and experimentally verifiable mathematical law through geometrical transposition. In 1590 Galileo had learned from his hydrostatic interpretation of the motion of falling bodies nothing more than qualitative lessons such as the progressive acceleration of motion. His interpretation even led to obviously false predictions, which he peremptorily declared to be in conformance with experience: light bodies—which, in view of their specific weight, contain less of the virtue of lightness than do massive bodies—must fall faster than the latter at the beginning of their fall. At this time Galileo had not yet been converted to the religion of facts. The *this must be thus* imperative still took precedence over the *this is.*

It is difficult to identify with any certainty the events that led him at last, at the beginning of the 1600s, to abandon the hydrostatic principle as the primary explanation of falling bodies. Most likely, while refining his experiments, he became aware of the considerable discrepancies between that which he *wished* nature would do and that which it actually did. In the Florence Library some of Galileo's notebooks were found containing apparently precise results of experiments on an inclined plane. But their exact dates, as well as their detailed interpretation, remain unsure.[3]

Still, in 1604 Galileo proposed another principle and announced at the same time that he held to the mathematical law of falling bodies that, he believed, geometrically followed from the new principle.

He announced this double discovery to his protector and friend, Paolo Sarpi, theologian of Venice:

> Reflecting on the problems of motion for which, for the demonstration of the accidents which I have observed, I lacked an utterly indubitable principle that I could take as an axiom, I have arrived at a proposition which is most natural and evident, and with it being assumed I can demonstrate the rest, namely, that *spaces traversed in natural motion are in the squared* [*doppia,* i.e., double] *proportion of the times* and consequently *the spaces traversed in equal times are as the odd numbers ab unitate,* etc. And the principle is this, that the natural moving body increases its speed in the proportion that it is distant from the beginning of its motion, as for example, in assuming that a heavy body is falling from the point *A* through the line *ABCD,* I suppose that the degree of speed which it has in *C* to the degree of speed it has in *B* is as the distance *CA* to the distance *BA,* and so consequently in *D* it will have a degree of speed more than in *C* according as the distance *DA* is more than *CA.*[4] (See fig. 4.)

The principle Galileo invoked in this letter is erroneous: velocity, in free-fall, is proportional not to the distance traversed but to time passed, which is a different matter. If velocity were proportional to the distance traversed, then that distance would increase exponentially with time and thus more quickly than in true motion, in which it increases only as the square of time. This error makes it even more remarkable, therefore, that Galileo was able to articulate the correct law of motion as a geometrical consequence of the principle; this law must have been suggested to him by another source, most probably through painstaking measurements. Erroneous as was the principle invoked, it nevertheless constituted a great step forward in relation to the analysis advanced in *De Motu,* since it does negate the Aristotelian principle that states that a constant cause can engender only constant effects. For the first time, in fact, a cumulative process for natural motion could be envisaged: degrees of acquired velocity are added to themselves. This is already an awkward suggestion of the principle of inertia (in a moving object isolated from outside influences, all motion acquired is acquired forever).

From that time on Galileo possessed the algorithm that would allow him to calculate, and thus to predict, the "accidents" of uniformly accelerated motion. He knew that projectiles follow a parabolic trajectory. He could calculate the inclination that would give to an artillery piece the required range and showed, in particular, that the maximum range is obtained at a 45° angle of fire. These figures delighted the captains of Venice and Florence, where he moved in 1610. Then, however, a newcomer from Holland called his attention to the remarkable properties of a certain arrangement

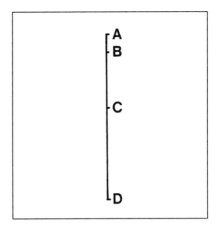

Figure 4. Galileo's law of falling bodies.

of optical lenses allowing distant objects to be seen as if they were near. From that time on he devoted all his ingenuity to exploring the skies and developing new arguments to support Copernicus's cosmological system.

■ Recognizing Time's Dominion

As he grew older, Galileo perfected the principles of the new science that his admiration for Archimedes, his faith in the mathematical intelligibility of nature, and his defiance of the arguments of authority had instilled in him. In his last work, the *Discourses on Two New Sciences,* he reiterated the correct law of falling bodies and demonstrated its essential relationship with the true principle of dynamics: in the motion of free-fall, velocity is proportional to the *time* elapsed, not to the distance traversed. The person of Sacredo is charged with asking the ingenuous question in this work: why choose a principle having to do with time, which seems complicated and abstract, rather than a more simple and concrete principle in which velocity could be proportional to the distance traveled? The response given by Galileo, in the third "day" of his book, is convoluted and inexact. The correct answer is, nevertheless, to be found in that same work: to understand uniformly accelerated motion, it is necessary that "we fix our attention upon the supreme affinity existing between motion and time." Time, and not space, is the hidden parameter of dynamics, the true wellspring of physics.

This is truly one of Galileo's greatest and most fundamental discoveries, the greatest doubtless in the field of dynamics, made at the culmination of a life consecrated to the search for a new understanding of nature: there is a supreme affinity between motion and time, that is, between the breath that courses through nature, which makes it vibrate, change, and live, and this tangible quantity whose signature is the ticking of the clock. The youthful error of Galileo, of which he was doubtless aware, was to place too much confidence in Archimedes, to force nature into certain forms and geometrical necessities, when dynamics actually obeys a logic other than that of form and space.

The invention of the "time" of dynamics then allowed the new science's founder to overthrow Aristotle's view, which defined time as the number of motion. The number of motion, according to Aristotle, exists only by virtue of motion, depends on motion, is only a reflection of this primary reality, and probably exists only insofar as it can be numbered, which is to say, enumerated by human intelligence, as Aristotle himself suggested in the *Physics*. Galileo showed to the contrary that there exists in nature something primary that rules motion and that he identified with the "time" of clocks, that of the water clock he invoked to describe his experiments on the inclined plane, that of the beating of his pulse or of the swaying of a chandelier in the cathedral of Pisa, and finally that of the mechanical clocks of Padua.

This change of view doubtless inspired in him speculations on the future developments of his discoveries: "And now we can say that the door is open, for the first time, to a new method already supported by many admirable findings which, in the years to come, will inspire mankind."[5]

At the same time we must recognize that, if the new theory of mechanics had indeed changed the landscape by assigning time a role as a primary reality, it did not provide a means of concretely apprehending this reality any more than did the ancient physics. All the clocks that I have just listed were in fact signs of time, mere imitations of it, with no guaranties that they would always be able to work synchronously. In one of those reversals and ironies common to the history of the sciences, the new mechanics outlined by Galileo, which would be perfected by his successors Gassendi, Descartes, and Newton, would soon produce a numeral of time. The numeral of the new mechanics' "time," however, is in fact identified with space through use of the principle of inertia: Two intervals of time are equal in size if, over the course of the second, a body moving free of any impinging force traverses the same distance as does the first.

4

Newton and the Discovery of the Law of Gravitation

In 1666 Isaac Newton was twenty-three years old. Some years earlier an epidemic of bubonic plague had forced him to leave Cambridge, where he was pursuing his studies at the university. Retiring to the county of his birth in Lincolnshire, he meditated on the great scientific questions of his time by concentrating his thoughts on them, waiting for the first glimmers of understanding to appear and slowly grow until the answers he sought stood brilliantly illuminated.

Among the great problems of his time was the development of a theory of mechanics. The law of inertia had been known for several years, thanks to the works of Galileo and his successors. According to this principle, material bodies isolated from all outside influence spontaneously preserve their state of motion. If they are at rest, they remain immobile. If they are in motion, they retain a rectilinear and uniform motion. For bodies subjected to an outside influence (after Newton this would be called "subjected to a force"), Galileo also came up with a special descriptive law, that of bodies subjected to gravity: they fall with a uniformly accelerated motion. If the body is thrown horizontally with a velocity v, its motion combines horizontal displacement at a uniform velocity, in conformance with the principle of inertia, with a vertical displacement at a uniform acceleration, in conformance with the law of falling bodies. Its trajectory describes a parabola.

While Descartes, on the Continent, tried to form a theory of the laws of mechanics and to understand them using acts of contact (direct shocks or, when needed, the intermediary of ethereal "vortices"), Newton sought to explain the world in another way, through the medium of immaterial influences and the qualitative explanations then in vogue in the field of alchemy. But first he had to put his ideas to the test. He could not use his notions of immaterial influences to formulate a theory of mechanics unless they truly explained the laws of motion and led to quantitative and verifiable predictions. Perhaps he had already decided on the term *force* for that immaterial influence, "the cause of true motion," which could not be attributed to the principle of inertia. But he needed proof.

In 1666 he was already aware of the explanatory power he could expect from the idea of force by applying it to the circular motion of the planets. These do not move in a uniformly rectilinear fashion; rather, they turn around the Sun. They are, therefore, subjected to a force. This must be comparable to what we call "centrifugal force," which makes taut the cord linking us to a stone we whirl around us, as in ancient slings. Simple mathematical reasoning showed him that, if the planets were actually subjected to such a force and if, in addition, their orbits obeyed the descriptive laws enunciated by Kepler, then this force must vary as the inverse of the square of the distance.[1] This is the famous law of universal attraction.

But this first rough intimation (Newton well knew that the motion of the planets was not uniform and that their distance from the Sun and their velocity varied throughout their orbits) did not yet provide the proof he sought. After all, saying that the planets describe their trajectories under the influence of a force that is directed toward the Sun and that varies inversely to the square of their distances from the Sun is only a more condensed expression of Kepler's laws. But what if this force also explained other mechanical phenomena, such as the falling of bodies?

It is possible that this fortunate inspiration came to Newton when he saw an apple fall from a tree during his strolls and his country musings in Woolsthorpe, since Newton himself reported it in his writings. It could also be, perhaps, a fiction he allowed to circulate when pressed by questions on the intellectual trigger that helped him perceive, in the depths of his contemplation, "the first glimmers of light that began slowly to shine." Impelled by a certain appetite for the sensational, Newton's biographers wanted to symbolize his genius by using the image of the stroke of inspiration that, at the fall of an apple, led him to understand universal gravitation, although this event was no more, at best, than the crystallization of that understanding.

■ The Law of Universal Gravitation

In Newton's time other thinkers were already striving to bridge the gap between celestial and terrestrial mechanics. Descartes wanted to use his "vortices" to explain celestial motion, as well as the motion of terrestrial projectiles. Hooke, a contemporary physicist and compatriot of Newton's, was working independently of him on similar ideas, but he was never able to support his theories with mathematical proofs. Newton's success is due not so much to his genius as to his stubbornness and the rigorous precision of his reasoning, which enabled him to demonstrate that the hypothesis of universal gravitation furnishes an explanation of the world system and to articulate the basis of a complete and precise theory of mechanics.

As early as 1666 Newton understood that the force holding the planets in their orbits around the Sun was the same as that holding bodies to the surface of the Earth or making them fall in accordance with the law of accelerated motion discovered by Galileo. This force is universal. It must, therefore, apply equally to the Moon. The Moon falls continually toward the Earth, he told himself, as a stone thrown horizontally describes a parabola. But, he reasoned, the Moon, carried away by its own momentum, then somehow misses the Earth and finds itself in the end always at the same distance from the Earth. This scenario can be verified by calculations. If the Moon "falls" toward the Earth while obeying the Galilean law of falling bodies, as would a stone thrown from a tower, its movement in relation to the horizontal plane must be a parabola, at least as long as the force of gravitation linking it to the Earth can be considered as having constant direction (fig. 5).

If the force of gravitation varies proportionally to the square of the distance of the object with respect to the center of the Earth, as Newton thought, then one should be able to deduce a relationship between the period of lunar revolution, the terrestrial radius, and the distance between the Earth and the Moon,[2] thanks to the equivalence proposed between the elementary arc of the circle of lunar motion and the elementary parabolic arc derived from the law of falling bodies.

Newton established this relationship and tried to verify whether the Moon obeys it. The period of the Moon's revolution around the Earth (the lunar month) was well known. The ratio between the Earth's distance from

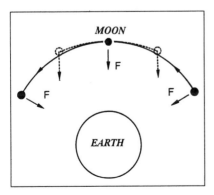

Figure 5. A key element in Newton's discovery of the law of universal gravitation came from demonstrating that each element of the circle of the Moon's revolution around the Earth could be thought of as a segment of the parabola (dotted line) characteristic of the Moon's fall toward Earth. The apparent difference between the two movements is due to the progressive change in orientation of terrestrial attraction.

the Moon and the terrestrial radius had been estimated using the lunar parallax, that is, the difference between the angles of the lunar attitude at the same moment as recorded in different terrestrial observatories. This ratio was estimated in his time to be 60, which was fairly accurate. Newton knew, of course, the value for gravitational acceleration at the surface of the Earth. The terrestrial radius, on the other hand, was less well known. Galileo had written, in his *Dialogue on the Two Systems of the World,* that the terrestrial radius was 3,500 "millaria," or Italian miles, about 5,300 of our kilometers. Newton probably used this value, although it was no more accurate than the ancient estimate of Eratosthenes. This value was, in fact, too small by almost 20 percent. The force holding the Moon in its orbit therefore appeared to be weaker than predicted by Newton's hypothesis of universal gravitation.

The difference was greater than he could accept, and the error, if error there was, could come from only one of two causes: either the value used for the terrestrial radius was false, or it was wrong to assume, as he did, that weight at the Earth's surface—due to the mutual attraction of all parts of the terrestrial globe—was the same as if all the Earth's mass were concentrated at its center. Assuming, of course, that the hypothesis of universal gravitation itself was not wrong . . .

It is important to note that at this time Newton did not yet know how to demonstrate, as he would do later on, that the gravitational attraction exerted by a large spherical body is the same as that of a pinpoint body of the same mass placed at the very center of the sphere. Perhaps it was this inability that paralyzed him. Since he did not suspect that the value of the terrestrial radius was false, he did not recognize the value of his discovery and put off pursuing his research on this theme.

This interlude would last, in fact, for twenty years. The invention of differential calculus enabled him to demonstrate, in 1679, that homogeneous spherical masses and pinpoint masses are equivalent from the point of view of gravitation. Weight at the surface of the Earth is therefore the same as if the entire mass of the latter were concentrated in its center. Some years later, after a long correspondence with Hooke and Halley, he demonstrated rigorously that the trajectory of bodies submitted to a force conforming to the law of universal attraction (in $1/r^2$) in relationship to a fixed center describes an ellipse whose center of attraction occupies one of the focal points. The calculations he made in 1666 to deduce the law of universal attraction from the elliptical orbits of the planets were not, therefore, only approximations, as he thought at the time, but rigorous results.

Finally, in June 1682, Newton learned at a meeting of the Royal Society that, acting under orders from the king of France, the astronomer Jean Picard had obtained, about ten years earlier, a precise measurement of the terrestrial radius through triangulation of the distances between Amiens and Paris. Picard's measurements gave the terrestrial radius a value equivalent to 6,370 kilometers, instead of Galileo's assessment of 5,300 kilometers that Newton had used. The new measurement was precise to the thousandth place, and there was therefore no doubt as to the significance of its difference from Galileo's figure.

Newton, history tells us, was staggered. He took up his calculations of 1666 using the new value for the terrestrial radius. His excitement was such that he had to get help with his calculations to arrive at the certainty that the Moon is kept in its orbit by a force that conforms exactly with the predictions of the law of gravitation.

In 1684, after twenty years of trials and reflection, the last details of his work were completed. Only some few friends were kept informed, among them Edmund Halley, who insisted that they be published. The manuscript *Principia Mathematica Philosophiae Naturalis* was written very quickly and presented to the Royal Society on April 28, 1686; the society decided to print it at its own cost. The work appeared in May 1687. For nearly fifty years it was ignored by continental physicists, who rejected the idea of attraction at distance, preferring the qualitative explanation of Descartes's vortices. It was Voltaire, followed by Laplace, who finally gave Newtonian science its due. It would soon thereafter be transformed into a sort of religion.

■ What Is a Force?

Newtonian science is built on the idea of force, "the cause of true motion"— or more precisely, the cause of true changes in the state of a body's motion (since, as we have seen, a body in rectilinear and uniform motion remains in this state without needing any force applied to it). But what is a "true" change according to Newton? It is not a change of motion as observed in relationship to a frame of reference of just any kind; rather, it is a change of motion in relationship to a preferred system of reference, which Newton called absolute space and time.

For example, imagine that a stone falls from a mast on the bridge of a ship advancing through the channel of a port. Its movement can be analyzed either in relationship to the lines of the deck and to the lines of the railing of the ship's bridge or in relationship to the netlike lines formed by the blocks of stone constituting the wharf. In the first instance, the motion of the fall would be defined as having a uniformly accelerated rectilinear motion; in the second, as describing a parabolic trajectory. In both cases, however, there is only one "true and absolute" force explaining the "true" change in the motion of the stone: gravitation, the cause of its vertical movement.

The horizontal movement seen when the wharf is used as a frame of reference describes only the displacement of the ship with respect to the wharf. It was in order to separate the true motion caused by a force from such relative displacement that Newton proposed, at the beginning of the *Principia,* the existence of a framework allowing the motions of physical bodies to be related to each other in an absolute manner. Absolute space-time, he explained, could not be confused with space and time in their everyday sense. Common space and time are linked to a particular frame of reference that might move imperceptibly with respect to absolute space (as long as the movement is rectilinear and of a constant velocity) or that might be displaced imperceptibly in time in relationship to absolute time.

Here, therefore, are the absolute space and time willed to us by Newton, which have become so profoundly anchored in our culture that it is difficult for us to see their conventional and arbitrary side. If the Newtonian concept of absolute space and time, infinite and existing independently of us and of material objects, was indeed a useful and even necessary step in constructing a theory of mechanics around the idea of force, it nonetheless was foreign to the physicists who preceded Newton; it would also be vehemently debated after him.[3]

As to the true nature of the force, Newton prudently professed his ignorance ("hypotheses non fingo") and affirmed his desire to content himself with describing its observable consequence, motion itself. Doubtless this was wise on his part. Today, the theory of quantum fields describes the force as the manifestation of incessant exchanges of virtual particles[4] of the "boson" class between material particles, or "fermions." One could reasonably allow that the ontological nature of the force remains a bit of a mystery . . .

■ The Infinitesimal Element of Time and Causality

In writing the laws of dynamics and in drawing on the concept of force, Newton advanced science toward a causal interpretation of time. The fundamental equation of dynamics, that acceleration is equal to the force imposed on a body divided by its mass, allows us to predict the state of a system at the next moment, infinitely close to the present one, on the basis of its present state. This differential calculation can be repeated a large number of times. The game of integral calculus (invented independently by both Newton and Leibniz) allows the state of a system at *any* instant to be determined from its present state.

The appearance of differential equations with respect to time in Newton's laws of dynamics is often considered to be the birth of causality in physical theory, since it translates into mathematical terms the idea that the future state of a system can be determined from its present state. This was, specifically, the opinion of Albert Einstein, who honored Newton as the great inventor of causality. The Newtonian idea of causality was still too abstract, however. Whether he wanted to or not, Einstein transformed it by raising it, in the theory of relativity, to the level of an operational principle; according to this principle, any disturbance of a physical system can have consequences for other physical systems only after time passes, thus allowing a signal to propagate.

What remains today of Newton's fundamental breakthrough? Modern life, our system of education founded on the requirements of punctuality, scholastic exercises on the charts of train schedules, geographic maps—all this inculcates in us, from childhood, a very Newtonian idea of space and time. This is why we have such difficulty perceiving the absurdity of questions such as "What lies beyond the limits of the universe?" or "What existed before the creation of the world—or before the Big Bang?" We marvel at the apparent modernness of Saint Augustine, who was already addressing similar questions fifteen centuries ago: "Time did not exist before heavens and earth."[5] But few among us know or have really assimilated the Kantian critique of the concepts of space and time. Kant constructed this critique specifically to chart the boundaries between knowledge and faith, to free science from metaphysical presuppositions, to deliver geometry from the shadow of theology to which Newton had in fact ascribed it.[6] For Kant, space and time are not things in themselves but "forms of intuition"—in other words, they constitute a canvas that allows us to decipher the existence of the world. According to Kant, things "in themselves" are neither in space nor in time. It is the human mind that, in the very act of perception, superimposes these categories, which are its own and without which perception would be impossible. This does not exactly mean that space and time are illusions or pure inventions of the human mind. These frameworks are imposed on us through empirical contact with nature and are not, therefore, "arbitrary." They no more belong to things in themselves than they belong to the mind alone; rather, they exist because of the dialogue between the mind and things. They are, in the final analysis, an unavoidable product of motion itself by means of which the mind searches to apprehend—to understand—the outside world.

Time and the Natural Sciences

While physicists were developing a concept of time that would facilitate their comprehension of physical phenomena, the temporal dimension of the natural sciences was also emerging from the shackles of mythology and religious dogmatism. The latter in particular relegated time to a reduced span: a literal reading of the Bible, then *de rigueur,* accorded the world only a few thousand years of existence. Breaking with this tradition, thinkers would recast the history of the world in terms of millions of years.

More than any other person, Charles Lyell was the author of this breakthrough in the Earth's history. Bringing to a culmination an underground school of thought going back to Aristotle and featuring such names as Leonardo da Vinci, Nicolas Sténon, René Descartes, James Hutton, and more, he recognized for the first time, in the texture of the Earth's surface, the results of a long, slow process stretching over hundreds of thousands of years; above all, he was the first to advance scientific arguments supporting this expansion of time scale.

In the realm of biology, the lengthening of the time scale was the work of Charles Darwin, contemporary, disciple, and friend of Lyell. Armed with the latter's writings, Darwin broke with the creationist tradition and Linné's purely morphological descriptions. He extended Buffon's first speculations on the evolution of animal species and replaced geography (as a principle of classification) with time. Finally, he developed the immense historical fresco that links us, by way of our cousins, the great apes, to the most primitive mammals and to the reptiles and fish.

Together, Lyell and Darwin overturned an entire worldview, one in which the Earth was only a stage setting hastily prepared for the coming of humankind. The late arrival of *Homo sapiens* on the scene and its modest place in the history of the universe did not sit well with a mythology of religious inspiration, imbued with a creator so terribly anthropomorphic in his impatience. Resistance was plentiful and stubborn.

■ Geological Time: From Leonardo da Vinci to Lyell

During the Renaissance, religious specialists in sacred texts estimated that the creation of the world might go as far back as about 7,000 years. The first thinkers who tried to understand the actual physiognomy of the globe, with its landscapes and its geological strata, thought that the Earth had been as they now saw it since its origin, either created thus in one swoop or shaped in the first phases of its existence by catastrophic upheavals, in particular by the Flood.

Leonardo da Vinci was a great forefather of geology, as well as of physics.[1] He demonstrated that, contrary to the generally accepted opinion of his time, marine fossils found in the ground of continental Italy cannot be vestiges of the Flood because they are found in different and superimposed geological layers. His observations launched stratigraphic analysis and descriptive geology. But if he suspected the considerable time spans implied by sedimentary formations, he did not pursue it in any systematic fashion and did not propose any coherent unifying theory.

Thomas Burnet, a man of the Church and a contemporary of Newton (with whom he maintained a scientific correspondence), wrote *A Sacred Theory of the Earth* in the 1680s. He was still, as regards time, in thrall to the ancient myth of circularity. He conceived of universal history as a "Great Year" that, passing through all the phases of a symmetrical cycle, began with Christ and would return to Christ.[2] According to Burnet, at the beginning of time the Earth was in a state of perfect harmony, its globe smooth and without blemish. It temporarily lost this peaceful appearance during one phase of catastrophes. The Flood marked the beginning of geological time, and the end of time will be marked with a fiery cataclysm before the Earth regains the perfect harmony that will presage Christ's triumphant return. Burnet's approach is still interesting, however, for it is endowed with the outlines of a preliminary sketch of *time-becoming;* in fact, he asserts that "it would in no ways suit with the divine wisdom and justice to bring upon the stage again these very scenes, and that very course of human affairs which it had so lately condemned and destroyed."[3]

It is James Hutton (1726–97) who gets the credit for having shown the futility of relying on a theory of catastrophism to justify the Earth's current physiognomy. He argued that if the known causes of erosion such as the sea, rain, wind, ice, volcanic activity, and localized earthquakes had extended over a sufficient amount of time, for periods incommensurably longer than those previously taken into account by the sacred writings, then the geology of the current era was perfectly understandable. Hutton's reflections certainly constituted a step beyond Burnet's, even though, curiously, Hutton rejected the idea of the direction of time. Only duration counted for him, and the history of the Earth is that of a battle without end between the plastic forces that raise mountains and the destructive forces that erode them.

Charles Lyell, who is generally considered to be the true founding father of modern geology, was another who did not appreciate time's arrow. He conceived of a grand fresco of geological history whose principles are lumped together under the name of uniformitarianism. He maintained that currently known geological forces are sufficient to fashion the most impressive of geological reliefs, such as the Grand Canyon of Colorado, if given enough time. In his determination to deny the thesis of catastrophism, however, Lyell wanted to see a blind process, with neither cause nor goal, in the formation of geological strata (to which he gave a nomenclature very close to that still in use today), a process therefore subject to sudden changes, to reversals as well as to forward evolution. He preached a perfect parallelism between the history of life and the history of the environment and thought that life-forms manifest a total adaptation to climatic conditions and to the reigning geology of their times. Denying the successive nature of geological transformation, he did not see why conditions extinct today might not recur and genera or even species that had disappeared with them could not thus return. One of his adversaries dedicated a biting cartoon to him, showing the "Professor Ichthyosaurus," symbolizing Lyell, leaning with commiseration over the vestiges of a disappeared and surely inferior species: *Homo sapiens* (fig. 6).

In the course of his geological excursions, Lyell came to understand all the consequences the presence of fossils held for the relative dating of geological layers. Because he did not believe in any type of arrow of time as regards the growth of the tree of life, the arguments drawn from the presence of fossils could not, however (according to him), be anything more than statistics: the presence or absence of this or that species had for him no temporal significance in itself, although the more distant in time two geological layers were, the smaller the proportion of fossils they would share.

But if Lyell was a truly important innovator, it was not only because he gave geologists a method by which to index the different geological epochs succeeding one another throughout the Earth's history. It is more because he understood, as did Galileo, the importance of time as an explanatory principle in the geological sciences. One of the incidents by which Lyell became aware of the immensity of the geological epochs and of the slow pace of their changes took place in 1828. That year Lyell visited the Auvergne; there he observed volcanoes and several sedimentary deposits. He noticed that the chalk deposits in one of them separated themselves into very thin layers, a bit like the annual striations of growth in tree

Figure 6. A caricature dated 1820 represents Charles Lyell as an Ichthyosaurus of the future commenting before his students on the discovery of a human skull. "You can see immediately that this skull belonged to an animal of an inferior order; the teeth are of an insignificant size, the power of the jaws is minimal, and overall it seems astonishing that this creature would be able to feed himself."

trunks. These layers are formed from the valves of a minuscule crustacean of the genus *Cypris,* and Lyell reasoned that each layer corresponded to a year's deposit. He counted about thirty layers in 2.5 cm of soil, which was piled up to more than 230 m. Thus, this terrain of relatively recent formation must have resulted from a continuous marine sedimentation over the course of about 276,000 years.

One can imagine that the difficulties then impeding the development of geology came about in large part from the inability to date geological layers. At that time only a relative chronology was possible. Let us not forget that methods of absolute dating depend on measurements of radioactivity, unknown until 1906. One can draw a parallel between the difficulties in elaborating mechanics, which was halted by the absence of the means to measure brief time spans precisely, and the tardy stammerings of geology, which had to await the twentieth century for instruments capable of measuring the very long time spans that characterize it.[4]

■ Time for the Living: From Buffon to Darwin

The analysis of time spans characteristic of the evolution of life is closely linked to that of geological eras. But the development of scientific thinking here was assembled in a relatively short time. The main act was played out between the publication of Linnaeus's *Systema Naturae* in 1735 and Darwin's *Origin of Species* in 1859.

Carolus Linnaeus typifies the partisan of creationism: his mind seems completely closed to the fascination of time's power. The great naturalist developed his work on taxonomy using admittedly painstaking but purely morphological observations, without any historical perspective.

His intellectual adversary in France was Georges de Buffon. Buffon preferred substituting biological criteria for the purely morphological criteria of Linnaeus. In particular, he defined and separated the species according to the nowadays well-accepted criteria of interfecundity. In addition, he tried to characterize the relationships between species, not only through morphological resemblances, but also by using the criterion of proximity of geographical habitats. Little by little, Buffon was led to recognize the role time played in biology; he admitted, for example, that variations within a species are due to the action of natural forces (climate, nourishment, environment). But he was not yet able to imagine the quasi-unlimited duration Darwin would claim for the action of these forces.

The true founder of transformism was the Marquis de Lamarck, later energetically opposed by Darwin. Lamarck incontestably placed time and its power at the heart of his philosophy. Nonetheless, if the history of science today calls Darwin rather than Lamarck the father of evolutionism, it is in part because Darwin, with his observations aboard the *Beagle,* collected a rich harvest of proofs and concrete arguments lacking to his elder, a man of solitary reflection and theoretical synthesis. In addition and above all, the scientific community did not regard Lamarck as truly one of their own, since instead of basing his conclusions only on past causes, he accepted the validity of a final great cause, an immanent force in nature oriented in time.

Darwin's greatest success was to argue the theory of evolution on a nonfinal basis, by invoking the perfectly blind forces of chance and natural selection. The success and renown of Lyell and Darwin, therefore, must be seen as more than a coincidence: both savagely fought the idea that the Earth and life have a destiny, and both recognized that chance could be the ultimate explanation for what might appear to be the workings of destiny. This formula is still very powerful today. The use of chance, however, to some extent hid the point at stake, for it would take better explanations than those generally presented today to give a reasonable account for chance's ability to simulate destiny.

In the field of biological sciences, Darwin brought time through a truly crucial stage. One of the moments of revelation came after the *Beagle's* voyage in the Pacific. At the Galapagos Islands Darwin suspected that the isolation of the bird colonies had led to genetic drift and the multiplication of cousin species. This observation, in conjunction with his reading of Lyell's *Principles of Geology,* inspired in him the concept of a time scale appropriate to these phenomena of genetic drift, imperceptible while they are at work in raw nature and accelerated by the practice of stock breeding. Finally, after his return to England in 1838, the *Essay on the Principle of Population* by Malthus helped Darwin crystallize the principles underlying the mechanisms of natural selection. In *The Origin of Species,* published in 1859, he clearly expressed his conviction that natural selection generally acts very slowly, over long intervals of time, and on only a few inhabitants of the same region. These slow, intermittent results, he believed, agree with the geological record of the rate and manner of changes in the world's inhabitants.

■ ■ ■

In summary, three traits seem to distinguish Lyell and Darwin from their predecessors. First, they shared an intimate conviction of the importance of *time* as explanatory parameter in the natural sciences. Secondly, they *explicitly refused* to take recourse in *final causes,* which they judged to be antiscientific. Contrary to widespread opinion, this attitude was not evidence of a polemical position vis-à-vis religion. When Darwin embarked on the *Beagle,* he secretly hoped to accommodate a religious interpretation of the history of life. At the end of his life he tended toward a nonmilitant agnosticism. A believer, Lyell thought for a very long time (until 1862) that humanity was apart from natural history, that it was a miraculous last-minute "addition."

Finally, both used not rhetorical discourse but precise examples to argue their convictions. This was the case for Lyell in the evaluation of the time scale of sedimentation as it was for Darwin in the study of the diversification of the bird species in the isolated colonies of the Pacific islands.

The absence of irrefutable means for precisely measuring the slow workings of time over the scale of biological and geological epochs was, as already noted, their true handicap. Yet in 1874, Ernst Haeckel, a fervent and quasi-fanatic partisan of Darwin (one can suspect that with him anti-religious motives influenced his attitude, independent of his scientific qualities), noted in *The History of Creation* that the history of the Earth must be measured in paleological and geological eras, each lasting thousands of years, if not millions or billions. Such spans of time are impossible to imagine, Haeckel commented, and geology lacked a sure mathematical basis to express even approximately in numbers the length of the unit of measurement that was needed.

Time and Space

Measuring space and time would seem to require different and unrelated techniques. The ancient Egyptians were highly skilled in the art of surveying using ropes and rulers, but they measured time only very roughly with sundials and water clocks. Even today, measuring distance using a graduated ruler seems simpler and more concrete than measuring an interval of time with a timepiece. From the viewpoint of the original motivations and epistemological significance of concepts, however, rather than from the viewpoint of technical practices, this preeminence of space over time does not hold up. Not only are measurements of space and time dependent on each other, but in many ways, time appears to be a primitive notion that must be understood before any attempt can be made to comprehend that of space.

For example, when the Greek astronomer Aristarchus of Samos, in the third century B.C., wanted to compare the size of the orbits of the Sun and the Moon to the dimensions of the Earth, he measured the apparent angular diameter of those bodies. We think that he did this by comparing the time the Sun and Moon take to rise and set on the horizon and the time the Moon takes to traverse the cone of the Earth's shadow during an eclipse.

Later on, the difficulties involved in measuring time and progress in geometry broke this link between space and time. The preeminence of space, established by geometers such as Euclid and Archimedes, lasted until the classical era but has since been assailed. Scientific developments not only agree with reflections on the practices of the ancient astronomers, who furnished us with only sparse findings, but also confirm the findings of experimental psychology, which has for some hundred years studied the development of the notion of time in children.

■ Primacy of Space or Primacy of Time?

It would be natural to expect that studying the development of the notion of time in children might provide us with the means to settle the question of the primacy of space or time once and for all, a question to which the historical approach offers only vague direction. Unfortunately, difficulties arise when we apply the tools of experimental psychology to the study of newborns and nursing infants. These can express their sensations in only a very imprecise fashion. Even so, general observations suggest that very young children are aware of time before they acquire any perception of space.

Thus, the cries of newborns are interpreted as a mark of their impatience to return to their mother's breast, since they stop crying on contact with it. On the basis of such observations, Jean-Marie Guyau had, as early as 1902, conceived of time as "being, in a certain way, the conscious gap between a need and its satisfaction, the distance between 'the goblet and the lips.'"[1] In 1928 Pierre Janet went even further:

> The concept of duration begins, in my opinion, with those very simple sensations, which appear with the first regulation of actions. They are different efforts. There are certain efforts: effort of continuation, effort of beginning, effort of ending, which exist from the moment a living being does something. . . . It is at that moment that the first sensation of duration appears. It perfects itself . . . through behavior already more precise: behavior of presence, of absence and above all the behavior of waiting.[2]

More recently Paul Fraisse suggested that "the most primitive feeling of duration, therefore, arises from a frustration of temporal origin: on the one hand the present moment does not afford us the satisfaction of our desires and on the other it refers us to a future hope (end of waiting or of the action begun)."[3] Jean Piaget affirms that in the beginning, "duration is confused with impressions of waiting and of effort, even with the unfolding of the act, lived internally. As such, it must certainly fill the child's universe in its entirety, since no distinction is yet possible between its internal world and the external universe."[4] But he soon notes that this lived duration is still far from being the common notion of time, since, he says, it comprises no real befores or afters. It is only much later that the child will understand time in its directionality, according to the definition given in the preamble of this work. Piaget notes that at the fourth stage (8–9 months to 11–12 months) the child becomes capable of seeking a lost object when he sees it hidden beneath a screen or when a screen is placed between the object and his regard, proof that he has understood that objects retain their identities and their permanence while being displaced in time. But these clear examples of temporal orientation already imply the concept of a spatial object. In this sense the acquisition of the concept of time becomes, at this stage, parallel to that of space.[5]

Admittedly, the analysis of primordial sensations remains based largely on fairly rough data, however probable they may be. These do indicate, however, an original primacy of time over space, succeeded by a much more intriguing situation in the life of the infant. Historically, science seems to have followed a fairly similar path. Since antiquity the difficulties of measuring time on the one hand and the progress of geometry on the other rapidly relegated time to a position as a subordinate of space (the instruments to measure time indicated, after all, only a change in the position of a pointer on a graduated scale.) Shortly thereafter the preeminence of space in the arts and sciences, reinforced by the development of geometry, became glaring. This state of affairs lasted until the Galilean revolution.

Paradoxically, this revolution, whose most significant accomplishment was to reintroduce time into the heart of theoretical reflection, for a while still accentuated the primacy of space. Time was, certainly, at the heart of the new theory. But how to measure it? How to assign it a measurable and quantifiable magnitude? The theory then responded: by means of the space traversed. Making distance covered the concrete sign of time's passage was accomplished using the totally new principle of inertia, which stipulates that in the absence of any force acting on it, a body will travel in a straight line over equal distances in equal time intervals.

In practice, of course, this principle has never been directly used to measure time. Neither Galileo nor his successors could imagine or construct clocks using the rectilinear movement of a test body, free of all force, which would allow time to be read by means of a pointer indicating, at each moment, the distance traveled by this body. Consider, however, this statement by Galileo: the trajectory of a ball falling from a table at a nonnull horizontal velocity describes a parabola. What can this mean except that the horizontal distance traveled is in fact a measurement of elapsed time, in light of which the uniform acceleration of vertical movement can be deduced? Galileo dreamed of and then sketched, and his successors were able to construct, clocks with pendulums, the successful operation of which implied the validity of the laws of dynamics in general and that of the principle of inertia in particular. To place time at the heart of dynamics, as Galileo did, was the decisive step needed to ensure the coherence and reliability of mechanical methods of measuring time. It also justified the ancient practice of measuring time by linking it to the rotation of the celestial sphere, the uniformity of this rotation also being dependent on the validity of the principle of inertia and the fundamental principles of mechanics.

■ A Case in Point: Taking Bearings at Sea

Among all the kinds of physical measurements, the determination of longitudes perfectly illustrates the intricacy of measuring space and time. Here time regains its original importance, because measurements of space become less immediate in the middle of the ocean: reference points in relation to outstanding features of the landscape and measurement of distances traveled using a surveyor's thread become impossible. For taking maritime bearings, the *quantity sought* is a space, and the *given* is a time.

Because the invention of the compass allowed mariners to dispense with coastal features, navigators at the end of the Middle Ages returned to the ancient usage, introduced in antiquity by Ptolemy, of marking geographic locations on maps by their latitude and longitude.[6]

The longitude is the angle that must be turned around the axis of the Earth when moving on a parallel to encounter the meridian of reference, that of Paris or Greenwich. It is also the angle the Earth must turn around its own axis so that the spot at which one finds oneself is brought back to the plane formed by the axis of the Earth and the Sun when it is noon at Paris or Greenwich. Measuring it therefore entails measuring the difference in true solar time between the actual point considered and the point of reference. How then to measure the time lag between these two points? The local time, synchronized to noon at the passage of the Sun on the meridian, must be compared to the time of the clock at the point of reference, either Paris or Greenwich.

The first method, proposed as early as the twelfth century, consisted of observing the Moon's position on the sphere of fixed stars and then referring to an almanac to check the corresponding hour at the point of reference. But this method (called the "method of lunar distances") could not pretend to offer more than a rough estimate, due to the apparent size of the lunar body and its relative proximity, both of which cause errors of parallax. A more efficient astronomical method was introduced around 1670 by Cassini, who used the universal clock furnished by the periodic occultations of one of Jupiter's satellites.

The first suggestion to use ship's clocks, synchronized to the clock of Greenwich or of Paris so that local time and the time of the meridian of reference could be compared whenever needed, is attributed to the clock maker Gemma Frisius. He wrote in 1553: "Before setting out set your watch exactly at the time of the country you are leaving. See that the watch does not stop en route. When you have moved a certain distance calculate the hour of this place with the astrolabe; compare this with that of your watch and you have the longitude."[7] But the accuracy of the clocks of that time did not allow these earlier astronomical methods to be bettered. Even with the introduction of the pendulum and the construction of the first marine timepieces by Huygens after 1660, the error in determining longitudes remained unacceptable: Holmes, a ship's captain, testing Huygens's pendulums in 1669 by traveling 5,500 km in every direction, still made an error of 150 km on the point, finding himself on his return in view of the Cape Verde Islands. It was no doubt this experiment that caused the seventeenth-century pilot Jean Gonduin of La Rochelle to prefer hourglasses as marine timekeepers.

In 1714 Newton, president of the Royal Society, convinced the English Parliament to underwrite a contest for a means of more precisely measuring longitude, with a prize of £20,000 to the inventor who could reduce the margin of error to thirty nautical miles (forty-eight kilometers). The Academy of Sciences in Paris adopted a similar resolution in 1718.

Results came quickly: progress was made on suspension, on escapement mechanisms, on the balance wheel, on the spring, and above all on corrections for variations in temperature. This race for technological development, in which John Harrison, Pierre Le Roy, and Ferdinand Berthoud distinguished themselves in the eighteenth century, encouraged the development of the principles behind marine chronometers capable of accurately determining longitude. The precision of the point surpassed one degree (sixty nautical miles) in 1732, a half-degree around 1760.

The race to perfect marine timepieces was thus the decisive factor in improving the ability of ships to locate themselves at sea. Today, marine quartz watches are so precise that the error (about two nautical miles) is limited no longer by the ability to determine the time of reference but rather by the determination of local time using sextants. This error, hardly a problem for merchant marine navigation, is unacceptable for aircraft. In principle, taking bearings is done in the same way on an aircraft as on a ship. However, this method is rarely used today. The positions of aircraft and ships are determined most often by using either a network of ground-based radioelectric wave transmitters whose geographical position is known exactly or satellites with on-board atomic clocks.[8]

■ Today's Technology Reaffirms the Primacy of Time

Modern methods of determining distance using time are considerably more precise than their predecessors. For today's technicians, time is more concrete than space, because clocks are more precise than graduated rulers.

Methods based on the use of transmitters still presuppose that the velocity of the signal's propagation is known. Measuring the time difference between a signal's transmission and its return after reflecting off the object whose distance is sought constitutes at one and the same time the principle of radar, when the signal transmitted is in the form of short waves, of measuring distance by way of laser, when the signal is in the form of visible waves, and of sonar, when it is a question of sound waves. Measuring the time difference between the transmission of a flash of light and its return after it reflects off an obstacle is the best method for precisely determining the obstacle's distance. Thanks to the combined use of lasers and atomic clocks, which allow us to make measurements close to a billionth of a second, we have made spectacular progress in the precision with which we can measure the distances of distant objects such as the surface of the Moon. The uncertainty about this distance, several tens of kilometers around 1950, was reduced to about 10 cm because of the installation of several reflectors on the lunar surface. As long as separate standards for distances and durations were being used, any uncertainty about measured distances was due less to uncertainty about the duration of the signal's travel time than to uncertainty about the speed of light.

This last difficulty was removed at the meeting of the General Conference of Weights and Measures in October 1983. To surmount the limitations linked to uncertainty about the value of speed of light, the conference decided to modify the International System of Units. This change fixed a conventional value for the speed of light, determined to be 299,792,458 m/s. Conversely, the conference decided to abandon all definitions of a standard of distance and to keep only the time standard, defined by a frequency of the radiation of cesium. Because of these decisions, the meter lost its status as fundamental unit of measurement and became, from then on, no more than the distance traveled by light in $\frac{1}{299,792,458}$ s.

The theoretical and practical consequences of these changes were very important. The natural unit of length is now not so much the meter, heir of the old tradition linking that standard to the terrestrial circumference or to the length of a bar of platinum, as it is the light-second or its multiple the light-year, long used by astronomers, which now pervades all realms of physics. Space and time being unequivocally linked thereafter through the intermediary of the physical constant c, the relative precision of measurements of distance became ipso facto as good as those of time.

This new system of measurement, however, raised a new difficulty: it prohibits, in principle, the possibility of examining a hypothetical variation in the velocity of light over the course of cosmological ages. Only if the laws of mechanics are valid can they guarantee the reliability of mechanisms with which to measure time, including atomic clocks. The reliability of the latter depends on the validity of the laws of quantum mechanics and on the stability of the energy levels associated with excited atomic states, whose durability is presented as axiomatic. In addition, the new definition of the meter presupposes the permanence of the value of the constant c. In order that these definitions retain their coherence over the ages, either the laws of mechanics and the energetic atomic levels on the one hand, and the speed of light on the other, must undergo no variation over large time spans; or these variations, if they exist, must remain constantly proportional to each other. In fact, however, these conditions, which guarantee the coherence of the definitions, are only postulates, admitted a priori. The consequences of an eventual fault in these postulates would be imperceptible on the human scale, of course, since before the decision of the Conference on Weights and Measures no variation of the speed of light had been observed. However, the adoption of these postulates (if they do not correspond to an eternal truth) could greatly constrain the interpretation of cosmological findings for the reconstruction of the history of the universe and the prediction of its destiny.

Part 2

The Relativistic Theory of Causal Time

The Speed of Light Is Finite

The ancients thought that light was propagated instantaneously. Hero of Alexandria, a Greek physicist of the first century A.D., was the first to address this question in greater depth. He insisted that the speed of light must be so great that it could not be measured. This prudent position did not stop almost all the savants of antiquity, the Middle Ages, and even the Renaissance from believing in the instantaneous propagation of light. It was not until a chance minute astronomical observation, conducted for technical reasons, that the Danish astronomer Olaus Rømer became persuaded of the finite and measurable nature of the speed of light's propagation. Vision being our principal means of communicating with the world, this discovery would profoundly influence the theory of knowledge. In fact, it opened the door to a revolution in our epistemological conception of time: instead of being a substance or a means of relating events to one another, time became first and foremost a way of relating events to the observer through the instrumentality of sight.

■ Looking for a Universal Clock

Determining a longitude, locating ships at sea, or finding the earthly coordinates of a place demands comparison of the local solar hour with the solar hour of a place of reference. Until reliable, transportable clocks became available, the hour of the meridian of reference was calculated by comparing the position of the Moon to those given in the tables of an almanac. But this method required too many fastidious corrections. In any case, nature furnishes a more precise method of comparing clocks at two distant observatories: the periodic occultation of Jupiter's satellites. When a doubt arose as to how precisely an observatory's clock kept the time of the meridian of reference, it was enough to note the time at which the occultation took place and compare it to the time at which the phenomenon was observed at Greenwich or Paris; the indications on the local clock then could be corrected if necessary.

Erasmus Bartholin, professor of physics at the University of Copenhagen, his disciple Rømer, and Jean Picard (the French academician who measured the length of the terrestrial meridian and thus affected the destiny of Newton's work) decided to use this new method to determine the precise geographic position of the observatory of Hveen in Denmark, once used by Tycho Brahe.

The satellite of Jupiter that they proposed to use as their astronomical clock was Io, the closest to the planet of the four satellites of Jupiter discovered by Galileo. Perfectly visible with a medium-power telescope, it plunges regularly (every forty-two hours, twenty-eight minutes, and thirty-six seconds) into the planet's cone of shadow. The three astronomers therefore undertook a series of observations of Io at Hveen. At the same time, in Paris, which at that time played the role later assigned to Greenwich, the French astronomer Cassini did the same. Each group noted the hours of the successive occultations according to their local clock, which was synchronized to noon when the Sun was at its highest.

Their observations seemed to show a great regularity in the occultation, or eclipse, of Io. It seemed, therefore, reasonable that a calendar (the ephemeris) be drawn up in the form of a table predicting the dates and the hours of the appearance of these phenomena at the observatory of reference. Thereafter it would no longer be necessary to consult that observatory; one would need only to refer to the ephemeris. Colbert charged the astronomers of the observatory of Paris with drawing up the tables in question, and Cassini invited Rømer to come join him there to pursue their observations.

So it was that in 1675 they discovered an irregularity in the interval measured between two successive eclipses: the period varies continually over the course of the year by a little more than two seconds in relationship to its average value. It is longer when the Earth is moving away from Jupiter and shorter when it is approaching Jupiter. In the times of conjunction or of opposition, when the Earth moves perpendicularly to the line that joins the two bodies, the period takes on its average value.

Rømer proposed that the 2.4 second delay in Io's eclipse when the Earth flees Jupiter represents the time taken by light to travel the distance crossed by the Earth between the two eclipses.[1]

This explanation was daring at a time when the teachings of Aristotle and Descartes exercised a profound influence on Parisian intellectual society. These taught, in effect, that light was instantaneously propagated. Impressed by these arguments of scholarly authority, Cassini refused to accept the young Rømer's intuition and continued vainly to seek other explanations.

In September 1676 Rømer announced to the incredulous members of the

Academy of Sciences that the eclipse of Io foreseen for the ninth of November of that same year would be ten minutes later than predicted by the tables of the ephemeris, which was constructed on the basis of observations made when Jupiter and the Sun were in opposition. Effectively, while Jupiter and the Earth distance themselves from each other, Io inexorably accumulates several seconds of delay, eclipse after eclipse, compared to the time listed in the ephemeris. By the ninth of November that cumulative delay did indeed reach the predicted ten minutes!

In addition, Rømer calculated that the light of Io would take twenty-two minutes (the total delay over a six-month period) to travel the diameter of the terrestrial orbit, two times the distance of the Earth to the Sun. This distance was beginning to become known precisely enough for Rømer to attempt to deduce the value of the speed of light: he calculated that light traveled about 40,000 leagues in one second, that is, 225,000 kilometers. This estimated value was too low, due primarily to an error committed in calculating the time taken by light to travel the diameter of the terrestrial orbit, which is actually only three-fourths of the value Rømer used.

The explanation he proposed for the phenomenon he had discovered with Cassini did not persuade the academy scholars. In addition to the force of tradition and the authority of the ancient and modern astronomers, there remained a difficulty in experimental method predicted by Descartes. If the speed of light is finite, then the direction in which the "fixed" stars could be observed should change over the course of the year. In fact, to obtain the "apparent" direction of a star, it would be necessary to compound the velocity vector of the light according to the star's "true" direction with the velocity of the Earth's displacement through space.[2] Because the velocity vector of the Earth changes direction every six months throughout its orbit (to regain the same value each year), the direction of observation of the stars, ordered in a fixed system of reference linked to the Sun, should vary periodically. In 1675 the astronomers of the academy did not understand that the value of this "aberration of the stars" was too small to be shown with the instruments at their disposal. Some astronomers, such as Cassini's son, tried to observe it, but all their efforts were in vain.

In 1725 the astronomer James Bradley demonstrated the existence of the aberration and measured it precisely enough, thanks to a large four-meter zenithal sector constructed for that purpose. Observations made with this instrument showed that the stars describe an annual ellipse on the celestial vault whose large axis covers an angular distance of forty seconds of arc.

Bradley's measurements not only confirmed the finite character of the speed of light but allowed it to be given a new determination, independent of the first: the size of the ellipse is directly linked to the time light takes to traverse the diameter of the terrestrial orbit. The value Bradley found for this time lapse was sixteen minutes, rather than the twenty-two indicated by Rømer, which led to a speed of light close to 300,000 km/s.

Finally convinced by this new proof, the Paris astronomers gave posthumously the homage and recognition due to Rømer: a commemorative plaque was dedicated at the site of his lecture.

■ ■ ■

In the eighteenth century, the discovery that the speed of light is finite had important consequences for conceptions of space and time. It was the first blow against Newton's absolute time, a first step toward a return to the preeminence of local time. A distinction was made between the date at which an event was perceived and that at which it was really produced, and popular astronomers began to speculate about time machines (machines that would allow travelers to go back in time). In the nineteenth century the most famous of these astronomers, Camille Flammarion, wrote in *Lumen* the fantastic story of a traveler thrown into space at the speed of 450,000 km/s. Armed with a powerful telescope, he sees events on Earth roll backward in time, exactly as in a film threaded in reverse. The theory of relativity would later show the inanity of such fantasies, any voyage at a speed greater than that of light being impossible. But the idea that measurements of space and time are relative, an idea based on the finite velocity of the propagation of light waves, was already to be found in the writings of this romanesque tradition launched by Rømer's discoveries.

8

Einstein and the Theory of Relativity

Time and space are linked not only by the way in which they are measured. Charting events using independent parameters of space and time, as proposed by Newton, was so arbitrary, especially after the discovery that the velocity of light is finite, that it was bound to be reexamined one day. This system of reference assumes that time is the same everywhere, that it runs uniformly for all and in all places, and that the measurements of distance between two points always give the same results, whatever the concrete conditions of their measurement. But what are the concrete conditions of measuring intervals of time and space? How can one be sure that time truly passes in the same way for all observers?

We have seen how first the invention and then the perfecting of clocks allowed a socialization, a putting-in-common, of lived duration. Where the experiences of everyday life passing in one place are concerned, a consensus as to the dates and times of the events under consideration is easy to obtain. But it is a different story where precise measurements are taken in the framework of a scientific experiment conducted by observers at a distance from and even in motion relative to each other.

It was Albert Einstein who reexamined the edifice established by Newton, a physical theory founded on decoupling Cartesian space from absolute and universal time. Einstein demonstrated the practical impossibility of synchronizing clocks if they are moving in any way at all with respect to each other. He thus showed the Newtonian definition of universal time to be arbitrary. He then turned to producing a precise operational definition for the notion of time and thus to redefining simultaneity using purely physical operations (reading a clock, measuring distances with a ruler, or transmitting signals of known wavelength at a known velocity). The new method of charting and locating events that resulted from this work opened the door to the development of a new edifice of physical theory, the theory of relativity.

■ Newtonian Time Reexamined

One story reports that, at the age of sixteen, while he was in school at Aarau, Einstein was already mulling over the problems posed by the limitation of the speed of light. He asked himself what would happen if he looked at himself in a mirror while traveling in a very rapid vessel. At this time, light was known to be an electromagnetic vibration traveling in a medium that was conceived of as being a subtle and massless stuff called *ether.* It was supposed that the waves scattered through the ether (at the velocity c) like ripples on the surface of a pond, which travel by way of successive disturbance of the water's surface. By moving very quickly through the ether, a body, Einstein thought, could catch up with and perhaps even pass the wavefront of emitted light. He therefore speculated that if the mirror were held up in the direction of the cosmic vessel's movement, the emitted light would take longer and longer to reach it as the velocity v of the vessel approached the value c. The image, thought Einstein, would become progressively dissociated from its source, in that it would represent the source not at the present instant but as it was earlier. If the vessel traveled at the speed of light relative to the resting ether, the image would inevitably disappear, since the vessel would then travel in concert with the light waves, which thus would be stationary with respect to the vessel.

Starting in October 1896 Einstein pursued his studies at the Polytechnicum of Zurich. Often forgoing the lectures of his professors, he studied on his own Maxwell's and Lorentz's works on electromagnetism. His reflections, which were often linked with his first schoolboy speculations, revolved around the problems with the theory when it was applied to bodies in motion. After many false starts, he decided that the key to resolving these difficulties lay in revising the Newtonian concept of time. This idea, according to his own recollections, came to him abruptly after conversations with his friend Michelangelo Besso during May 1905: "It came at last to my mind that time was suspect . . . time is not defined in an absolute manner but in inseparable connection with the velocity of the propagation of signals."[1] Thus, following in Galileo's footsteps, Einstein rediscovered the concept of time's fundamental importance in physical theory.

A particularly explicit thought sequence inspired Einstein to denounce the Newtonian concept of absolute time as arbitrary (fig. 7). Imagine a train that runs an iron track at a certain speed. Two lights flash on the track at two separate points. "Common sense" would suggest the following criterion for simultaneity: the two flashes are simultaneous if they are perceived at the same instant from a point O_1 situated midway between the devices that produced them. An observer on the track may easily determine that the two flashes are simultaneous for him by using this definition. Now note

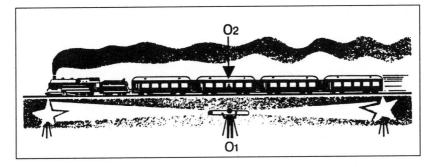

Figure 7. A device installed alongside a train track allows two simultaneous flashes to be generated during the train's passage. An observer (O_1) situated on the ground between the two sources and armed with a periscope allowing a view in both directions simultaneously would perceive them at the same time. However, the two flashes would not be simultaneous for an observer (O_2) sitting in the center of the moving train.

which two compartments of the train are exactly opposite the devices at the moment that they flash. Suppose that O_2 is the point on the train marking the midpoint between these two compartments. Would a traveler sitting in O_2 also see the two flashes at the same instant? No: he would see the flash occurring in the direction in which the train is traveling before seeing the flash occurring behind the train, since he is moving along with the train to approach the former and conversely fleeing the latter while its light is being propagated.

Thus, the finitude of the speed of light waves—which we use to date distant events—makes it impossible to adopt an unambiguous definition of simultaneity. The Newtonian proposition of an absolute time that is the same for all systems of reference is clearly unwarranted.

■ About the Electrodynamics of Moving Bodies

It is important to note that this radical criticism of the concept of time did not result from a direct frontal assault by Einstein but was the outcome of his in-depth examination of a technical subject: certain paradoxical aspects of the electrodynamics of moving bodies. According to classical theory, an electrical current induced in a moving conductor can have either an electrical or a magnetic cause, depending on which system of coordinates is chosen to describe it. Reflection on this problem led Einstein to question Newtonian time, in the same way that Galileo was led to criticize the geometric model of physical theory through his reflections on the phenomenon of falling bodies.

During his stay at the Polytechnicum of Zurich, a reading of August

Foppl's book entitled *Introduction to Maxwell's Theory of Electricity* greatly influenced the direction of Einstein's thinking. He was astonished at the dissymmetry that Maxwell's theory allowed when dealing with a current induced in a conductor in motion relative to a magnetic field.

Take a copper ring and a magnet. Whether one moves the copper ring in front of the stationary magnet or conversely holds the magnet and moves it in front of the stationary copper ring, an "induced" current appears in the copper ring. This current has, of course, the same value in both instances as long as the orientations and the relative velocities of the two objects are the same. According to Maxwell's theory, however, the explanation for this phenomenon is different in the two instances. His theory contends that induced current has two distinct causes, according to whether the conductor is moving in a fixed magnetic field (in which case the current is attributed to deflection, by the magnetic field, of the charged particles present in the conductor and in motion with it) or the magnet responsible for the magnetic field is moving in front of the motionless copper ring (in which case the current is attributed to the appearance of an induced electrical field acting on the fixed charges present in the conductor). The theory predicts that the current has the same value in the two cases, but this result appears to be a fortuitous coincidence in the effects of two quite dissimilar causes.

The theory of relativity resolves this paradox, which Maxwell's theory leaves unexplained. Magnetic fields and electrical fields are considered not as two different values of nature but as the projections or particular points of view of one value: the electromagnetic field. In the same way in which the apparent length and width of an object depends on the angle from which it is viewed, the magnetic field and the electrical field depend on the arbitrary reference system chosen and whether the conducting body is at rest or in motion relative to it.

■ The Two Basic Postulates of Relativity

The significance of Einstein's thoughts about the finite value of the speed of light, the movements of the "ether" in relation to the Earth and to the sources of light, and the paradoxical aspects of electrodynamics lay in the unification of these elements (electromagnetism and the principle of relativity of motion) into one coherent theoretical system.

By opening the door to a new definition of time in its relationship with space, Einstein immediately became aware that he could preserve Maxwell's equations in their exact form, fixing the laws of electromagnetism and predicting the constancy of the speed of light in all the frameworks of inertia. He then considered this property as an inviolable result and posed the two following postulates as fundamental principles of any future physical theory:

The postulate of relativity: The Newtonian notion of absolute rest—in other words, of rest in relationship to an absolute space—is without foundation, for it corresponds to nothing observable. Electrodynamic or mechanical phenomena have no properties corresponding to this idea. The notion of "ether," required for conceptualizing absolute rest in the case of electromagnetic waves, is superfluous and can therefore be abandoned.

The postulate of the constancy of the speed of light: The speed of light is independent of the motion of its source, as explicitly predicted by Maxwell's theory.

These two postulates should, in the framework of Newtonian absolute time, be incompatible with one another. Indeed, the Galilean law of the composition of velocities indicates that, if the velocity of the source measured in the system with the "ether" at rest (system at absolute rest) is v, then the velocity of the emitted light, measured under the same conditions, must be $c + v$. This rule, used implicitly, leads to the curious results foreseen by the adolescent Einstein when he imagined himself looking into a mirror in a very fast vessel. The reconciliation of the two preceding postulates thus demands a revision of the basic principles of mechanics—a revision possible only as long as the choice Newton used for charting events in space and time is arbitrary and can be modified appropriately. In fact, the two postulates impose the form that must be taken by this modification in a very precise and unique way, codified in the "Lorentz transformations."

The postulate of relativity itself, to which Einstein's theory owes its name, is the first of the two postulates expressed above. It is in fact a natural extension of the postulate of Galilean relativity, according to which one cannot, with any mechanical experiment whatsoever, discern the motion of the laboratory in which this experiment is taking place if this motion is rectilinear and uniform in relation to "absolute space" (usually represented by a system of reference linked to the fixed stars). The postulate of relativity extends the preceding statement to all the experiments of physics, whether mechanical or electromagnetic.

■ Time Dilation and Length Contraction

As soon as time loses its absolute character, the measurement of an interval of time between two events can depend on the system of coordinates used as reference. As a direct consequence of Lorentz transformations, neither temporal nor spatial measurements are the same when taken from an inertial frame of reference (linked to the fixed stars) and from a second frame of reference in uniform, rectilinear motion relative to the former.

The measured interval of time separating two events taking place successively at the same place is greater in the second frame of reference: this is the dilation of duration. This remarkable property was experimentally verified for the first time only in 1938, when it was demonstrated that terrestrial laboratories observe unstable particles at rest to decay much sooner than these same particles transported by cosmic rays (at velocities close to the speed of light).

In addition, distances on objects at rest in the first frame of reference are longer when measured in that frame than when they are measured in the second: this is the contraction of lengths, which explains the negative results of experiments like Michelson and Morley's.[2] In his article of June 1905 in the review *Annalen der Physik*, Einstein exposed the paradox of the electrodynamics of conductors in motion and criticized the concept of simultaneity in Newtonian mechanics. He presented the postulates of relativity and demonstrated the formulas of Lorentz transformations of coordinates,[3] applying them to the dilation of duration and to the contraction of distance, as well as to the impossibility of a body attaining the speed of light. Mechanics and electrodynamics were reconciled, and the way was open for the announcement of the equivalence between mass and energy some months later:[4] $E = mc^2$.

■ Propagated Causality

In the article of June 1905 Einstein announced his findings in a form that seemed to limit their epistemological import. He challenged the concept of absolute rest but did not yet state that the form of the equations of physics must stay the same in all the reference frames of inertia, as he would later write. He affirmed the constancy of the speed of light but made no reference to a limiting velocity of energy transfer in the void or of information transfer. He limited himself to a technical presentation of results obtained on the basis of the two postulates of relativity, especially in the context of the electrodynamics of bodies in motion.

In truth, it seems that in 1905 Einstein did not yet perceive the true nature of the revolution he had just initiated. The critique of the concept of simultaneity rests on the propagation of light waves, and the postulate of relativity places the propagation of light at the center of physical theory. But why make the speed of light play such a role in the description of nature? What is the origin of the alleged despotism, reinforced a posteriori by the announced results that imply the speed of light to be a limiting velocity no material body can attain, because it would take an infinite amount of energy to attain it? Can it be said that what seem to be all our experiences are mediated by vision and the use of light rays? This is an insufficient explanation. A blind population would certainly develop a physics similar to ours. Another basis must be found for the preeminence of lightspeed.

Many epistemologists consider Einstein to have introduced in 1905 a physical theory whose touchstone is the paradigm of causality propagated at velocity *c*. In the theory of relativity, time is a parameter that allows events united by a causal chain, or susceptible to being so united, to be labeled in successive order. The versatility of the different ways of charting events in time and space in accord with Lorentz transformations derives from this absolute order regulated by propagated causality. The theory of relativity is therefore a *causal theory of time.*

In fact, it was not Einstein who took the first steps down this road but Hermann Minkowski, his former professor. In 1907 Minkowski published an article entitled "Fundamental Equations of the Electromagnetic Phenomena of Bodies in Motion," and the following year he gave a resounding speech entitled "Space and Time," in which he announced provocatively, "Sirs, the ideas of space and time I would like to develop before you have their roots in experimental physics. . . . Henceforth, space as such and time as such must be considered as dead, and only the union of these two factors retains its validity."[5]

After having formally introduced four-dimensional space-time (where the fourth dimension is time converted into space using the conversion constant *c*), Minkowski divided this into zones called "past," "future," and "elsewhere." Only the events of the past are perceived by an observer situated at the origin, who can exercise influence only on events located in the future. The observer can neither know nor influence the events taking place in the elsewhere (fig. 8).

Minkowski showed next that, if spatial and temporal measurements vary when going from an inertial system of coordinates to another system, one quantity defined as the "invariant interval"[6] remains constant despite the change. The invariant interval between two event-points located on the tra-

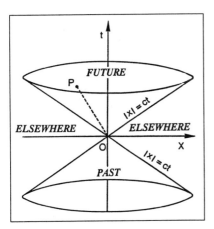

Figure 8. Minkowski's relativistic space-time. The vertical axis represents time, and the horizontal axis represents only the coordinate *x* (the coordinates *y* and *z* are not represented). The observer is supposed to be situated at the zero point and at time *t* = 0. For this observer, the separation of relativistic space-time into past, future, and elsewhere is delimited by the "light cone" whose trace on the plane shown is given by *x* = *ct*. The separation between the zero point and the event *P* of coordinates *x, y, z, t* is measured by their "invariant interval." For the points situated on the light cone, this invariant interval is null. In the usual way, the light cone is also represented in perspective in tridimensional projection (*x, y, t*).

jectory of the same ray of light is null. For an observer, the event-points situated in the past or in the future are at a negative invariant interval (squared). Those of the "elsewhere" are situated at a positive invariant interval. Thus, in relativity theory, the use of light rays as possible carriers of causal influences determines the structure of the universe.

Einstein took a stand on the question of causality in 1905, some time after the publication of his article but before Minkowski proposed his generalization. This is shown by an exchange of correspondence with Philipp Frank, who had just published an article entitled "Causal Law and Experience": "[Einstein] . . . agreed with me," Frank noted on this subject, "that, whatever may happen in nature, one can never prove that a violation of the law of causality has taken place. One can always introduce by convention a terminology by which this law is saved. But it could happen that in this way our language and terminology might become highly complicated and cumbersome."[7]

The theory of relativity safeguards the principle of causality. It is, moreover, the only adequate answer in the case of a causality propagated in a vacuum at velocity c. This is the reason certain authors today substitute another formulation for the second postulate of relativity as it is found in Einstein's article of 1905: they replace the speed of light in a vacuum with the velocity of the propagation of information-bearing signals. On the one hand, this substitution generalizes the original formulation, insofar as one can imagine propagated actions that would be transmitted by means other than light waves (gravitational waves for example). On the other hand, propagation of an energy at a velocity higher than the speed of light is acceptable, as long as this propagation concerns phenomena not susceptible to mediating the propagation of a cause.

These past few years there has been a great deal of talk about the "nonseparability" of the products of atomic decay, which implies instantaneous transmission at a distance of an *influence* able to modify statistical observations made of many such decays. But it has been shown that this phenomenon cannot serve to propagate *information* at a velocity superior to that of light.[8] More generally, an entire class of interpretations ("stochastic interpretations") of motion in quantum mechanics has the same implication. In each case, an element of randomness in the observed phenomenon precludes using it to transmit information in a way that would violate the principle of propagated causality.

■ ■ ■

Hans Reichenbach is without a doubt the philosopher of science who has made the best study of the theory of relativity as a causal theory of time. In his book *The Philosophy of Space and Time,* he remarks that "relativistic physicists have indeed formulated a correct theory of time, but they have left their opponents in the dark concerning the epistemological grounds of their assumptions."[9] In another work, he was more specific:

> Einstein's principle of the limiting character of the velocity of light is based both on negative and on positive evidence. It is negative evidence for the principle that no signals faster than light have ever been observed. But Einstein's principle would never have been accepted, had it not been based, in addition, on positive evidence. The special theory of relativity would lead to absurd consequences if the principle were not true; the time order of causal processes could be reversed, infinite amounts of energy could be produced, and so on. Consequences of this kind appear so unlikely that they supply an argument in reverse: it appears very unlikely that the principle of the limiting character of light velocity is false.[10]

Thus the theory of relativity enshrines time as parameter of causality. In this sense, it is a direct continuation of Newton's work, making time the parameter of immediate causality in the differential equations of dynamics. But the theory of relativity also raises questions about choosing systems of axes of coordinates. The results obtained up until 1905 were valid only for inertial systems of reference, that is to say, axes of coordinates pointed toward the fixed stars. But is it possible to develop a physical theory beyond this constraint? What would become of the principle of propagated causality in this new theory? Is the general theory of relativity, which as we know addresses this new situation, also a causal theory of time?

9

Time and the

Dance of the Galaxies

Just as with Newton's laws of mechanics, the laws of relativity developed by Einstein in 1905 cannot be applied unless an appropriate system of coordinates is chosen. For most applications, a system of coordinates whose origin is located at the center of the Sun and whose axes point toward the fixed stars represents just about the ideal system. A system of coordinates rigidly linked to the Earth and its rotation allows the laws of classical or relativistic mechanics to be applied only approximately. The differences are already tangible even at the level of many daily phenomena, such as the trajectory of an artillery shell or atmospheric circulation in meteorology. Einstein was well aware of this weakness. No sooner were the 1905 laws of relativity established than he dedicated himself to their generalization to any system of coordinates whatsoever. To do this, he needed first to understand in what way these other systems might differ from the systems valid under special relativity. The theory he developed between 1907 and 1917, called the theory of general relativity, again reinforced the idea that a clock could not function independently of the physical conditions of its environment. Combined with astronomical observations, it also allowed an attempt to reconstruct the history of the universe, with an enlarged temporal horizon henceforth numbered in billions of years.

■ The Sphere of Fixed Stars: A Special Frame of Reference

After the publication of the basic article on special relativity, Einstein returned to his work on its extension. He was inspired in this undertaking by Ernst Mach, who was also considering the special role of inertial frames.[1] The main characteristic of these systems is that they are oriented to the fixed stars, even if these are very distant. But why are Newton's laws valid only in systems of reference whose axes are firmly attached to the great masses of the universe? The explanation suggested by Mach is that these masses are in themselves the cause of inertia. In other words, a body's tendency to stay in a constant state of motion in an inertial system is a consequence of the action exerted on that body by the great masses of the universe.

■ The Equivalency of Weight and Inert Mass

Fictitious forces exist in a frame of coordinates in nonuniform motion relative to a system of inertia.[2] Specifically, if a body is at rest in a frame of inertia, it will seem to be subjected to a fictitious and constant acceleration when compared to a system in uniformly accelerated rectilinear motion. In 1907 Einstein understood that the force of gravitation, which imparts a uniformly accelerated motion to all terrestrial bodies, could be considered a fictitious force, insofar as it is nullified by the correct choice of the frame of coordinates. Einstein suggested the comparison with an elevator in free-fall, in which the occupants no longer feel the force of gravity. Today this situation is routine for the astronauts inside space stations orbiting the Earth. There is no gravitational effect in a system in free-fall.[3] The implications of such a statement are considerable. Could gravitation, this mysterious force on whose origins Newton "did not want to risk a hypothesis," be no more than a simple geometric effect, depending on the choice of the system of coordinates? For Einstein, the field of gravitation had only a relative existence, comparable to the electrical field produced by magnetic induction. (Here he refers to the phenomenon that led him to the theory of special relativity.)

This parallel, which suggests the relative character of gravitational forces, is possible only because of the equality of masses subject to inertia and those subject to gravitation—seen experimentally but until then unexplained. In a frame of coordinates linked to the Earth, the tension that attracts any body toward the planet's center can be measured. This tension is called the body's weight. In a frame of coordinates subjected to a vertical acceleration (e.g., a rocket) the resulting force on the bodies can be measured: for example, the pressure (force of inertia) exerted by a body on the floor of the rocket. For each body, weight and force of inertia remain rigorously proportional, so that it would be impossible to decide, through physical experiments inside a closed and windowless room, whether this room were accelerating in empty space or at rest near a massive body whose gravitational field was attracting it. One proof of the equality of weighted and inertial masses was furnished by the "plumb-line" experiment. Regardless of the material constituting the instrument's mass, the plumb line indicates the same vertical in the same place. But the tension of the line in fact results from two forces—a force of attraction, proportional to the weight of the lead, and a centrifugal force imparted by the Earth's rotation. The direction of the latter with respect to the vertical forms an angle equal to the site's latitude, and its intensity is proportional to the inertial mass of the body. If the relationship of the weight to the inertial mass depended on the material used, different plumb lines would indicate different verticals.

Thus, for Einstein, the equality of inertial mass and mass subject to gravitation, far from being a fortuitous circumstance, took on the value of a principle, that of the relativity of gravitation. It allowed him to interpret gravitation as a geometric effect. Of course, doing this required that the geometry be generalized to include, in the definition of the distances between bodies, an element that would reflect the presence of sources of gravitational fields.

The concrete elaboration of this theory, in which geometry is determined by the sources of the gravitational fields, required a great deal of work. Marcel Grossmann, an old friend of Einstein, mastered the techniques of calculating in non-Euclidian geometries. Both collaborated in applying them to the new relativistic theory of gravitation that Einstein was trying to construct.[4] Over the course of four years (1911–15) Einstein published the first articles on the general theory of relativity; the principle of equivalence was reviewed in detail in 1911, the notations of Riemannian geometry were introduced in an article written in collaboration with Grossmann in 1913, and the general equations of gravitation were presented to the Prussian Academy of Sciences in 1915. These equations allowed the laws of physics to be written using a uniformly accelerated frame or any other frame; they are in agreement with Einstein's principles, according to which "if we are going to do away completely with the difficult question as to the objective reason for the preference of certain systems of coordinates, then we must allow the use of arbitrarily moving systems of coordinates."[5]

■ Practical Consequences on the Behavior of Clocks

In the general theory of relativity, time retains exactly the same status as it enjoys in special relativity: it is a parameter of causality linked to the propagation of light. In the immediate vicinity of a point in four-dimensional space-time, a system of coordinates can be chosen such that the metric to be applied—the prescription for measuring distances considering the sources of gravitation present—would be that of special relativity; specifically, light would travel in a straight line with the velocity c.

Over great distances and in the presence of massive objects, however, light follows a trajectory that does not correspond to a straight line in Euclidian space. Thus, the rays emitted by an object located behind the Sun and grazing its edge deviate by a certain amount. Predicted in 1913, this deviation was verified in 1919 with the aid of a total solar eclipse. As with intervals of space, intervals of time lose the property of congruence (superposability when moved) characteristic of Euclidian geometry. In the neighborhood of massive bodies, or during acceleration, clocks slow down.

In 1915 Einstein found himself therefore in possession of the general principles and fundamental equations of the relativistic theory of gravitation. As does special relativity, this theory rejects the existence of an absolute frame of reference for space and time (the absolute Euclidian space and Newton's absolute time). In addition, it affirms that one can chose whatever metric allows the laws of physics to take on as simple an expression as possible, as long as this choice is experimentally confirmed. In general relativity, the trajectory of bodies free of all force other than that of universal gravitation is represented by a "geodesic," the natural extension of the straight lines of Euclidian geometry. The originality of the theory comes from eliminating gravitational attraction from the "true" forces of nature. It becomes thus a pure geometric effect. The new theory also predicts some remarkable effects, such as the curvature of light rays in the vicinity of the Sun and the advance of the perihelion of the planet Mercury (an effect already known at the time, but unexplained in the framework of Newton's theory).

■ Einstein-Mach: Harmonies and Discords

During this period Einstein wrote to Mach several times, notably in June 1913, crediting Mach's research as having inspired him. But Mach did not unequivocally support the theory of relativity. In fact, although he was, like Einstein, a true partisan of a relational interpretation of time and space (in which these two entities have neither reality nor meaning in themselves but express only relationships between objects), he was also an ardent positivist who would not allow himself to speak of the reality of objects and events outside of perception. He certainly felt some distrust of Einstein's convictions about realism. For Mach, scientific laws were no more than an economical means of simultaneously describing a great number of observations, and their "reality" had no other measure or meaning than in this degree of economy. The theory of relativity, however, whether special or general, gives to the paradigm of causality, expressed by the propagation of light at a finite and constant velocity, a "reality" that Mach thought to be unwarranted. Einstein, despite the warm funeral eulogy he gave to him in 1916, henceforth kept his distance from Mach and his principle.

■ Einstein Launches the Era of Modern Cosmology

The new perspective on the geometric order of the world introduced by general relativity allowed a novel vision of cosmology and the history of the universe. Einstein recognized that and wrote, in 1917, a new article entitled "Cosmological Considerations in the Theory of General Relativity."

In this article Einstein examined the solutions of the equation of general relativity (which links geometrical metrics on the one hand and the distribution of mass and energy on the other) in view of the ideal limit of a *homogeneous* and *isotropic* world, in which the small, accidental accumulations of matter, such as the planets, the stars, and so on, can be "averaged out" so that the only important fact is the mean density of matter in the universe. In other words, he took the point of view of an external observer situated at a great distance from the universe, one who is not concerned with such unimportant details as the existence of the Earth or the Sun. This generalization is in fact legitimate; after all, a light ray grazing the edge of the Sun undergoes gravitational deviation of less than two seconds of arc.

Einstein studied the possible solutions for his equation, taking into account estimations on the average density of the universe. He discovered that not one of the solutions led to a *stable* universe. How could it be otherwise, in a world in which the only constraints of geometry (the gravitational forces) tend to coagulate fluid material and to make any cloud of matter collapse on itself? There existed only two ways of escaping the conclusion of such a general collapse: counterbalancing the tendency toward gravitational contraction with a pressure that would tend to disperse the universal fluid, as if it had undergone an initial explosion, or admitting a weak repulsion between bodies, insignificant at the local scale but sufficient, at the scale of the universe, to counteract universal attraction.

Einstein did not hesitate: he chose the second alternative and proposed introducing a new term for repulsion of a geometrical nature, which he baptized the "cosmological constant," into the equation of general relativity. The cosmological constant would allow the stability of the universe to be preserved and the problem of its origin to be ignored.

The first alternative, however, did not escape Aleksander Friedmann's attention. In 1922 he published an article entitled "On the Curvature of Space," in which he emphasized the possibility of solving the cosmological problem without recourse to the term of repulsion. In this solution the universe cannot be stable; it must be expanding or contracting. If the universe is expanding, the radiation emitted by celestial objects should exhibit a reduction in frequency through the "Doppler effect" (a phenomenon of light waves similar to one of sound waves in which car horns traveling rapidly away from us seem to lower their pitch).

■ Astronomical Discoveries and Modeling the History of the Universe

While Einstein and a handful of theoreticians devoted themselves to the cosmological models suggested by general relativity, astronomical observation made new progress thanks to the American Edwin Powel Hubble and his large diameter (2.50 m) telescope installed on Mount Wilson. A specialist in nebulae, Hubble made three great discoveries, one after the other.

He was able for the first time to distinguish individual stars in certain great nebulae, proving that they are in fact galaxies similar to but outside our own. On October 5, 1923, he observed and photographed an individual star of variable brightness in the large nebula of Andromeda. The absolute brilliance of this type of star is calculable, and the measurement of its magnitude as seen from the Earth therefore allowed a direct estimation of its distance. Soon twelve stars of this type were identified in the same galaxy, and Hubble could announce that it was situated about 900,000 light-years[6] from ours, the diameter of which is only about 100,000 light-years.

Hubble's second observation concerned the number and the distribution of the galaxies. Counting the galaxies he could identify in a given portion of the sky with the help of the large Mount Wilson telescope, he realized that their distribution was globally isotropic: galaxies and groups of galaxies are found in all directions. Better yet, thanks to the measures of distance that he was then able to broaden to about 250 million light-years, Hubble was able to verify the approximate uniformity of the distribution of matter in space, and from there he was able to make a first estimate on the average density of the universe (about 10^{-30}g/cm^3), a parameter crucial to all isotropic and homogeneous cosmological models.

Hubble's third discovery is his most famous. Studying the luminous spectra of about thirty galaxies whose distances he had estimated fairly precisely (up to 6 million light-years), he noticed a systematic shift of the emission spectra toward the red, as if these objects were undergoing a Doppler effect due to their flight away from the Earth. He noted that the velocity of this flight, as deduced from the spectral shift, was roughly proportional to the distance of the galaxies; this was Hubble's law, researched from 1924 and published in 1929. Today the proportional constant between distance and velocity is called Hubble's constant. Its inverse has the dimensions of a duration characteristic of cosmological history; in effect, it is the time galaxies would have taken to disperse to their present positions in the universe if their velocity was always what it is today. Hubble estimated this time at about 2 billion years.

These results did not escape the attention of another relativity theorist,

Father Georges Lemaître, who in 1927 published an article entitled "Model of the Homogeneous Universe, of Constant Mass and Growing Radius, in Accordance with the Radial Velocity of the Extra-Galactic Nebulae." He proposed as a model for the history of the universe the explosion of an initial nucleus, a "primitive atom," possibly as large as our solar system. The date of this initial explosion was estimated using Hubble's characteristic time. Lemaître's suggestion was largely ignored by scientific circles until the astronomer Arthur Eddington became interested in it in the 1930s.

Around 1932 Einstein, convinced by accumulating evidence on the flight of the galaxies, admitted the superfluity of the cosmological constant he had introduced in his 1917 model to preserve the stability of the universe. He then rallied to the "Big Bang" theory and accepted the idea of a "birth," of a singular point in the history of the universe.[7] He would declare later that the cosmological constant was the greatest mistake he had ever made.

Curiously, for the past several years, theorists have begun once again to invoke the possibility of a nonnull cosmological constant during the period of "inflation."[8] This inflation, however, would have applied only in the very first instants of the Big Bang, giving way then to the regular expansion that has presided over the destiny of the universe for the past 15 billion years.

■ Cosmological Cooling: The Source of History

Hubble's law, which concerns the proportionality between the velocity of flight and the distance of the galaxies, constitutes a fine example of successful scientific induction. Although the distances accessible to astronomers are now twentyfold larger thanks to modern radio telescopes, the consequent extrapolation has not changed the law's essential proposition of a linear dependence between distance and velocity of flight. This perfect proportionality is a comfort to us when we think of the universe as surging forth from an initial singularity in the form of a gigantic explosion.

Such a statement, however, could not stand alone. Prudently, Hubble always refused to speak of the flight of the galaxies and held himself to the term "red shift," imitating Newton's "hypothesis non fingo." Today, however, new confirmation of the Big Bang model comes from unexpected areas: relativistic thermodynamics and elementary particle physics give weight to the hypothesis of an initial explosion from a nucleus of reduced dimension. In addition, observations in the field of particle physics indicate that the initial catastrophe might have taken place at a much faster rate than previously thought, from a universe with a very high energy density and a tiny spatial dimension.

Relativistic thermodynamics allows us to understand how the expansion of the universe could engender, and continue to engender, a profound thermodynamic disequilibrium between matter and radiation. This disequilibrium constitutes the true driving force behind the world's organization into differentiated structures; it ineluctably evokes Aristotle's prime mover charged with maintaining the motion of the spheres. The disequilibrium has two main phases. In a first phase, the expansion of the universe by itself causes the "adiabatic" cooling[9] of the fluid of which it is constituted. *But this cooling does not affect matter and radiation in the same way.* The equations of general relativity show, in fact, that the temperature of matter decreases as the inverse squared of the "radius" of the universe, while the temperature of radiation decreases only as the inverse of that radius. Thus, a permanent source of negentropy (the possibility of organization) is created in the universe through the coupling processes between matter and radiation. The role of gravitation must be noted among these coupling processes. In a universe of cool material particles, gravitational collapse results in the birth of hot stars. This creation of hot stars is the second phase of the organization of the universe into differentiated structures. And it is from a nearby star, our Sun, that terrestrial life and humanity draw all the negentropy necessary to them, now as before.

The first phase has a verifiable physical consequence of major importance. At the beginning of the expansion, the density of energy in the universe (first in the form of radiation, then of radiation and matter) was very high. Every photon emitted by a particle of matter was reabsorbed almost instantaneously by another particle of matter. Its energy served to heat this particle—the processes of emission and absorption counterbalance the differential cooling and maintain a thermal equilibrium between matter and radiation. During the expansion, however, the dimensions of the universe grow large enough that the densities and temperatures of matter and radiation cross a threshold below which newly emitted photons have practically no chance of being reabsorbed. They then continue to cool independently of matter. These photons wander through the universe flying in all directions, forming a homogeneous gas. They constitute an observable radiation, even in regions where telescopes cannot discern any particular source. This phenomenon is called "background radiation." The formation of these photons can be calculated as occurring at about a million years after instant 0, when the characteristic dimensions of the universe were a thousand times smaller than they are today and when the characteristic temperature of the photons was close to 1,000°K. Due to the continuous cooling they have undergone since then, their characteristic temperature today is no higher than about 3°K. This background radiation, predicted by the physicist G. A. Gamow in 1948, was discovered accidentally by the radio astronomers Arno Penzias and Robert Wilson in 1965. The form of its spectrum coincides with that of the theoretical spectrum of "black body radiation," which has a characteristic temperature of 2.7°K, testifying to its thermodynamic origin. The intensity of this radiation is constant in all directions, with relative variations no higher than one part in a thousand.

■ Inflation and the Time Barrier

Elementary particle physics gives information about the history of the universe at times even closer to that initial instant. Most of the light elements present in the universe, such as hydrogen, deuterium, and helium, could not have been formed other than at this distant epoch, since their formation demands very elevated temperatures on the order of 10^{27}°K. Their current relative abundance reflects the conditions that prevailed then. They confirm that the universe must have known a period of temperatures higher than 10^{27}°K.

We now have, therefore, three independent bits of evidence indicating a cosmological history of the relativistic "Big Bang" type: Hubble's law, background radiation, and the relative abundance of the light elements. Other scenarios that have been proposed do not account for observations made in one or the other of these three areas. Today almost all astrophysicists accept this model of the history of the universe, even if many uncertainties remain about the details of the birth.

Until 1983 most astronomers conjectured that the universe issued from a pinpoint singularity of null size and infinite temperature. According to the scenario proposed, a tiny fraction of a second after its birth, the universe grew to a micron, while the temperature was still higher than 10^{32}°K. At such a temperature, no form of matter or even of light is stable; this primordial universe cannot be represented other than as an empty space in which "quantas" continually appear and disappear almost instantaneously. Since 1983, however, some discrepancies linked to the extraordinary homogeneity of background radiation have led certain physicists to see the origin of the universe in a slightly different light, even more accelerated, although the temperature could have remained finite, so that it is still difficult to make pronouncements on its dimensions. The history of the universe disappears behind this explosive, catastrophic phase, as behind a heavy tapestry. During this "inflationary" phase, the size of the universe is thought to have grown with a velocity superior to that of light, which does not contradict the principles of relativity, since matter did not yet have a stable existence. In the new, modified scenario of the Big Bang, matter would have appeared during the course of partial reheatings in a universe already relatively spread out.[10]

Whatever the case, the laws of physics do not allow us to glimpse conditions before this initial phase. They present it to us as an insurmountable barrier to history beyond which time makes an abrupt appearance. Beyond this initial quasi singularity exists neither specific direction nor structure whose traces might be followed to reconstruct that history. Time becomes dilute, disappears. Once again Saint Augustine's question can be asked: "What was there before time existed?" Nothing, at least if the definition of time given in the preamble is accepted. Nothing, since time does not exist without an object on which it can act and leave its mark. This degree of freedom vanishes at the very moment when the notion of "object" is lost in the primitive emptiness.

■ The Creation of the Universe, Science, and Philosophy

The question of what the universe was before the beginning of time shows just how difficult it is to get rid of the Newtonian idea of a time in and of itself, which would be something like Clarke's "sensorium Dei," since it would have had to exist even before the world existed. For general relativity, the spatio-temporal structure is determined by the momentum-energy tensor, and neither time nor space can exist apart from this tensor.

The history of modern cosmology merits a book in itself (see specifically the works of Jacques Merleau-Ponty). Unlike so many other scientific discoveries, however, the idea of an initial instant in the universe does not constitute a scientific revolution, one of those epistemological breaks dear to Bachelard. The idea of the birth of the universe is no stranger than that of our own births. In fact, due to the practical impossibility of retracing time beyond the initial instants of either birth, the two are related in our minds. This is doubtless why most religions, well before the sciences, already accept it through the myth of the Creation.

The biblical story of the Creation in Genesis thus contains propositions that seem astonishingly contemporary, notably the idea that the universe had a beginning. And the primacy of light is also evoked: "Let there be light!" The biblical text does even better, since indeed the light seems infused in a preexistent nothingness, agreeing with the most recent ideas on the nature and the role of nothingness in high-energy physics: "In the beginning God created the Heavens and the Earth. The Earth was a formless wasteland and darkness covered the abyss." Certainly, however, the most modern question about the origin of time is not addressed. For that, it was necessary to await Saint Augustine's admirable texts ("time, it is You who created it, there was no time before time was . . ."), al-Ghazali, and Maimonides, who one after the other attacked Aristotle and introduced the idea of an absolute origin of time coinciding with the creation of the world. The central idea of modern scientific cosmology is therefore neither new nor in opposition to common sense. On the contrary, it displays a reassuring similarity between the history of the world and individual history, perhaps even a connivance or a conformity of destiny.

But why therefore did the Enlightenment philosophers who introduced the history of the solar system into astronomical science turn away from the idea of an absolute origin? Did they want to free themselves of the myths of religion in the name of rationalism? The fact is that Laplace, in his *Exposition du système du monde,* excluded all discussion on the origins of nebulae, for he refused all recourse to final causes, which, as he stated in *The System of the World,* are but imagined for the sole purpose of calming our disquietude on the origin of things that interest us. In the *Encyclopédie* of Diderot, Laplace's contemporary d'Alembert explicitly repudiates the idea that matter could not be eternal. And Spinoza, almost a century before Kant and Laplace, also did not want to hear about a history of the universe. It is clear indeed that the very profound immanence of Spinoza's God, allied with the infinity of his attributes, makes difficult the idea of a creation of matter, of an absolute beginning of matter, at least under the form of "natura naturans"—that is, in its profound essence, which is extraneous to either space or time. That would come back to admitting that God could exist outside nature, which Spinoza denied. Given Spinoza's philosophical influence on Einstein, it can appear that he was, with Mach, an important inspiration of the Einsteinian synthesis, up to and including the episode of the cosmological constant. Both sustained Einstein in the most illuminating flights of his genius, but both participated also in the few false steps he took.

■ ■ ■

Physicists and astronomers of this century progressively brought together the elements of what today resembles a proof, an experimental finding: the universe is expanding.[11] It was born 15 billion years ago from an initial, formidable explosion, the "Big Bang." Its destiny, infinite dilution in an infinite space or contraction toward a final singularity (the "Big Crunch"), poses another problem, for it depends strictly on the average density of matter in the universe, a subject about which a large uncertainty still exists. Time has an origin, but we do not yet know whether it has an end.

The Paradox of the Twins and the Story of the Time Machine

The development of general relativity reinforced the belief that a clock's operation depends on its physical environment. As early as 1905, when Einstein's original treatise on special relativity was published, it contained a section in the appendix in which he pointed out "a peculiar consequence" of Lorentz's equations of transformation, which regulate the change of the coordinates of space and time between two frames of reference in uniform motion with respect to each other. The section concerned the discrepancy that develops between the times given by clocks attached to the two frames of reference. Einstein indicated that this discrepancy, which is due to the dilation of duration, would not be nullified if the motion halted or reversed itself, even if the two frames of reference once again coincided. In relativity any acquired time lag is acquired forever. Each object henceforth would have to be associated with its own time, whose rate of flow would depend on its acceleration and the field of gravitation to which it is subjected. Of course, at everyday velocities, the time shift is so small that it cannot be marked with ordinary clocks. Recently, however, the precision of atomic clocks allowed it to be measured.

The prediction of a discrepancy between two clocks moving relative to each other violates common sense, which dictates that time should pass at a rate unaffected by any spatial movement an object might undergo. Another surprise result of the experiment, which at first appeared paradoxical, was the dissymmetry of the predicted results. The traveling clock should *slow* in relationship to the stationary clock, and the stationary clock should *advance* in relationship to the moving clock, despite the fact that one of the great principles of physics is the complete symmetry of rectilinear and uniform motion. During the process of going or returning, the moving clock could be said to distance itself from or to approach the stationary clock, but it could also be said that the latter (and with it the entire laboratory) distances itself from or nears the clock in motion. The principle of relativity stipulates that no experiment of physics may choose between the two definitions as long as it is a question of rectilinear and uniform motion.

Paul Langevin, professor at the Collège de France and friend and admirer of Einstein, was doubtless the first to propose the most imaginative and at the same time the most convincing explanation for this phenomenon.

■ Langevin Explains the Paradox

In a conference given in 1911 at Boulogne, Langevin illustrated the paradoxical aspects of time in relativistic mechanics by recounting a story of two twins, one leaving for a very high speed interstellar voyage while the other remains on the Earth. After 2 years the traveling twin returns to the Earth to find it aged by 200 years (the text of this conference was later published).[1] Langevin gave a particularly clear and pertinent explanation for the difference in the rates of aging that can be reread with pleasure.

> Suppose that two fragments of matter first meet, separate, then find each other again. We can be sure that observers linked to each during the separation will not have evaluated the duration of the separation in the same manner, that they will not have aged at the same rate. As a result, those having aged least will be those whose motion during the separation was the least uniform, who were subjected to the most acceleration.
>
> This observation furnishes the means, to anyone who would like to devote two years of his life to it, of knowing what would become of the Earth in two hundred years, of exploring the future of the Earth by taking a leap forward that would last two centuries in the life of the Earth but only two years for the voyager, and this without hope of return, without the possibility of coming to inform us of the result of the voyage, since all attempts of this kind would only transport him further and further ahead.
>
> This enterprise requires only that our voyager consent to being enclosed in a projectile fired from the Earth with a velocity close enough, although still inferior, to that of light, which is physically possible. The voyage would have to be arranged so that a meeting, with a star for example, would take place at the end of a year in the voyager's life and send the projectile back toward the Earth at the same velocity. Back on the Earth, having aged two years, the voyager will leave the vessel to find our globe aged by two hundred years, assuming that the projectile's velocity remained less than light speed by only one twenty-thousandth during that voyage. The most established experimental facts of physics enable us to assert that this would actually be the case.

It is amusing to speculate on how our explorer and the Earth would perceive each other's lives if they could, by light signals or by wireless telegraph, stay in constant communication during their separation, and to understand thus how the dissymmetry is possible between the two measurements of the duration of their separation.

While they are moving away from each other at a velocity close to that of light, each one will see the other flee before the electromagnetic or luminous signals sent to them, in such a way that it would take a very long time to receive the signals emitted during a given time. The calculations show that each of them would see the other live two hundred times slower than usual. During the year in which they moved away from each other, the explorer would receive from the Earth news covering only the first two days after his departure; during this year the Earth will be seen to perform the movements of two days. . . .

During the return phase conditions would be reversed. Each party would see the other live a singularly accelerated life, two hundred times faster than normal, and during the year it would take to return, the explorer would see the Earth pass through the events of two centuries. Thus, on returning, the voyager would find the world aged by two hundred years. . . .

To understand the dissymmetry, one must note that the Earth would take two centuries to receive the signals sent by the explorer during the time the projectile moves away from the Earth, which for him would last one year. The Earth would watch the voyager live, during this stage in his voyage, a life slowed down two hundred times, making the gestures of a year. During these two centuries the Earth would have to use an antenna two hundred times longer than the projectile's in order to receive the hertzian signals sent back. At the end of these two centuries the Earth would receive the news of the vessel's encounter with the star, which would mark the beginning of its return. In the two following days it would take to return, the Earth would see the voyager live two hundred times faster than normal, experiencing the events of only another year and aging by only two years. During these two last days of the voyage, the Earth would have to use an antenna two hundred times shorter than the voyager's antenna.

The dissymmetry is due to the fact that only the voyager has been subjected, at the middle of the voyage, to an acceleration that changed the direction of the projectile's velocity and brought it back to its point of departure on the Earth. It is manifested by the fact that the voyager sees the Earth move away from and approach equally over one-year intervals, while the Earth, informed of this acceleration only by the arrival of light waves, sees the voyager move away over the course of two centuries and come back during two days, over a period forty thousand times shorter.

With this premonitory text, Langevin clearly pinpoints the true cause of the dissymmetry in the rate of aging: the acceleration to which the voyager is subjected, which is not relative. General relativity, which was being developed at the time of Langevin's conference, could not help but support this point of view. A detailed analysis of this type of experiment within the framework of general relativity allows the value of the divergence between these two clocks to be worked out, a divergence matching that predicted by Langevin's story of the twins.

■ An Effect of Perspective?

The total renunciation of time as a self-consistent concept independent of objects and the constraints to which they are submitted is the surest road toward correctly interpreting the adventure of Langevin's twins. It remains, however, difficult to break the habit of thinking of time as a universal invariant. For this reason many thinkers who accepted Einstein's theory of relativity and the associated equations of the transformation of space and time coordinates nevertheless tried to construct a different interpretation that, because the traveling clocks or twins are brought back *at rest* beside their earthly counterparts, would not imply a divergence between the two clocks or the unequal aging of the two twins.

Bergson in particular held this opinion. To him duration as it is lived is an *immediate given of consciousness,* the mark of intelligent consciousness and the foundation of organized perception. He did not deny the pertinence of relativity theory, but he sought to give it a philosophical reading in conformance with his convictions on the central importance of the self. In the case of Langevin's twins, he distinguished real time, that of the immobile observer, from "apparent" time. The observer attributes apparent time to events inside the rocket because of an effect of perspective caused by the rocket's motion as it recedes or approaches.

Lorentz transformations give the rate of passing time (the duration of a clock beat) in a reference frame in rectilinear and uniform motion, as it is measured on a clock in the frame at rest. It can then legitimately be said that *seen* from the system at rest, the clock in motion slows in relationship to the stationary clock. But does it slow down, or does it *seem* to slow? Bergson argued that, as long as the clocks are not compared side by side and at rest, "universal time" passes at a rate unaffected by motion, and the effect yielded by Lorentz's equations is only one of perspective. As long as Langevin's voyager moves away, the twin, watching the clock the voyager took on the projectile, *sees it* tick more slowly in relationship to his own. But, according to Bergson, this effect of perspective will disappear at the moment that the space vehicle comes to a halt, so that the two twins will again be the same age. According to Bergson, the effect is comparable to the illusion that results from plunging a stick obliquely into a pond. The stick appears bent because the light refracts in the water. The illusion ceases once the stick is brought back into the homogeneous environment of the air.

Bergson's explanation, which marries Lorentz's equations to universal time, draws logically on a relativism of accelerations. This relativism of accelerations is contrary to experimental fact and does not stand up to a detailed analysis of the experiment within the framework of general relativity.

■ The Theory Tested by Experiment

The experiment imagined by Langevin invokes uniform motions that do not truly allow the traveling clock to be returned to a state of rest on Earth. When Einstein went to Paris in 1922, he and Bergson began a dispute that for decades would continue to engage partisans on each side. As recently as 1971 an American physicist published an analysis of the experiment within the framework of general relativity, in the terms of which the "apparent" slowing would be nullified if the traveling clock were effectively brought back to a state of rest on Earth.[2]

However, relativity theory's growing popularity, in addition to the first experimental verifications of its consequences in areas other than the dilation of time, fascinated the public. Science-fiction novels and films illustrated this effect, and fantastic "time machines" abounded. But most of these works contain a gross error. The heroes, tired of their temporal explorations, came back to describe to their friends what they had seen. If the theory does indeed allow the future to be explored—to know what would become of the Earth in 10, 20, or 200 years—it does not posit any means of returning, of coming back through time.

Saint Thomas shows us what best convinces doubters: to see the divergence of two clocks is to believe it. Historically, the complete experimental demonstration of unequal aging was accomplished in stages.

The lifetimes of unstable particles transported by cosmic rays at speeds approaching the speed of light have been measured since 1938. The average lifetime before decay of particles of known speed, as measured by the distance of flight, is several times higher than the average "at rest" life of these same particles, in the exact proportion predicted by the theory.

These measurements did not dissuade Bergson's supporters from their faith in an effect of perspective, since the particles used as clocks were "watched" *during* their motion. This criticism is applicable also to the series of experiments conducted during the sixties using circular particle accelerators, where researchers measured the lifetime of unstable particles (μ mesons) circulating in a ring with a constant speed close to that of light. In this case, the particles returned periodically to their departure points, so that the moving "clocks" and the clock of reference could, in principle, be compared. The experiment showed once again, with better precision, a dilation of lifespans conforming to relativistic prediction. But the moving "clocks" remained continuously in motion and were not compared *at rest* after their trip with the laboratory's clock. Apart from the better precision of the measurement, the experiment's value consists in the ease of analyzing the behavior of the clocks moving in a uniform circular motion, either in terms of special relativity (the calculated rate of dilation of the durations is then linked to the tangential velocity of the particles in their circular orbit) or in terms of general relativity (the delay in the clock's time is then interpreted as being linked to the centripetal acceleration of the circular motion).

Finally, in 1971, the American physicist Joseph Hafele, judging that atomic clocks were now sufficiently precise to allow observation of unequal aging, conducted the decisive experiment. Despite being "atomic," the clocks he used were nevertheless clocks, with a numerical dial giving readings down to a billionth of a second. Hafele envisioned a voyage around the world in an ordinary commercial airplane. The very limited time difference predicted would be expressed in ten-millionths of a second, but the precision and reliability of the timekeepers would leave no doubt as to its reality.

Four atomic clocks were put on the flight and then compared to four others that remained on the ground. The experiment was conducted twice. One voyage went east around the world, and the clocks were compared on their return to their departure point. A second voyage then took the clocks to the west. The time differences measured over the two voyages should not have been the same, since the Earth rotates toward the east and its own movement must be compounded with that of the clocks.

At the end of the westward voyage, the clocks on board were observed to *advance* on the clocks remaining on the ground by 273 billionths of a second (with an uncertainty factor of only 7 billionths of a second). This finding seems surprising at first, since relativity theory predicts that a clock in motion must slow in relation to a stationary clock; but the paradox is only one of appearances. Seen from a standpoint linked to the fixed stars, the clocks on board the airplane were at one and the same time in a less-intense gravitational field (due to their altitude) and a system turning less rapidly than the Earth (the speed of the airplane must be subtracted from the speed of the Earth). The terrestrial clocks therefore should have been slower than the transported clocks. The predicted divergence[3] conformed with the observation when the velocities and the altitudes registered during the voyage and their levels of uncertainty were taken into account. Relativity had triumphed once again.

On the eastward flight, the traveling clocks were part of a system whose rotation was more rapid than that of the Earth in relation to a reference point linked to the fixed stars. But they were also, due to the altitude of the flight, in a less-intense gravitational field. These two effects were in opposition, but the first carried the day. The traveling clocks, on their return to earth, were *slower* than the earthbound clocks. The difference in time they noted was entirely attributable to the effects foreseen by relativity. After the publication of these results,[4] no one further contested the existence of the relativistic effect of unequal aging.

■ **The Time of the Monads**

It might seem that abandoning Newton's absolute time because of relativity theory and the equivalence of spatial and temporal variables in measuring distance or duration had reopened the question of time as a fundamental concept in describing nature. That is not the case. Paradoxically, the concept of time acquires even more importance in relativity theory than it had in Newton's system. The paradigm of causality proposed by relativity obliges us to substitute for the immaterial time of Cartesian coordinates and the theological time of Newton an individual time attached to each object and to each being. This time marks such objects' reality, their becoming, the time of the clocks that rule the functioning of their internal wheels—in short, their *own* time. The equivalence of spatial and temporal variables does not alter the specificity of time, for time acquires all its value once it is situated in the system of reference where the object or the being *is*.

By the same token, this time, because it results from the coordination of the entire universe through the principles of relativity, also marks the presence of each thing and of each being in the world, their participation in the entire cosmological becoming, in a universe marked by the passage of its history. It marks the interdependence of their destinies and reminds us in this way of the Leibnizian concept of the "monad" applied to an individual and solitary thinking unit, which would be at the same time the reflection of the entire world.

The Limits of Causal Time

The theory of relativity allowed the ordering principle of propagated causality to be discerned, and its power is such that it led its supporters to equate causality with time. But the causal interpretation of time is burdened with a double and serious defect.

On the one hand, identifying time with causality provides a twisted image of time, for it deprives time of its arrow: causal time is no longer directional.[1] An attempt was then made to introduce another principle into the theory (unfortunately also often called the principle of causality) that would restore time's directional orientation by declaring that causality should be applied only in a single direction—from "causes" to "effects." But this proviso seems artificial and perhaps useless when viewed in relation to the laws of electromagnetism.

In addition, causality's field of application is perhaps not as universal as Einstein thought. In fact, it seems to apply only to macroscopic phenomena and not to the elements of reality in itself, as proven by the properties of inseparability demonstrated by certain interactions between elementary particles. The impossibility of describing separately the properties of quantum objects, even when the objects are at a large distance from each other, has long been among the principles of quantum mechanics, but only a recent series of experiments succeeded in clearly demonstrating it.

The need to turn to a subsidiary principle to restore past-future asymmetry and the inseparability of certain quantum properties seem to limit the universality of the relativistic principle of causality and to ruin any possible identification between the primitive notion of time and that principle.

■ Causality and Reality

The intuitive concept of time was refined during the development of physical theory (classical mechanics and relativity). It became a geometric concept allowing the positions of events on the world stage, deployed in its four dimensions, to be quantitatively fixed in concert with space. Correlatively, the concept of energy and the paradigm of causality appeared, both closely related to time.

The concept of energy is related to time by a principle of symmetry according to which the laws of physics are eternal and invariable over the course of time. In sum, this principle of symmetry stipulates that any physical experiment whatsoever must yield the same results whether the experiment is conducted today or tomorrow. Therefore, the energy must be conserved. The link between the law of the conservation of energy and the irrelevance of placement in time was formulated in 1860 by Emmy Noether.

The paradigm of causality is also linked to the conservation of energy and to time. According to this paradigm, a precise knowledge of the physical conditions of a system at any given instant ineluctably fixes its future. The development of relativity made this clear: causes habitually propagate at the speed of light, but never faster than lightspeed.

The theory of relativity and the analysis of causality taught us two new and essential things about time. First, contrary to Newton's convictions, all observers do not necessarily order events in space and time in the same way. The *time* that passes is, therefore, not one of the world's shared experiences. The relativistic concept of "one's own" time (called "proper" time) reinforces our impressions of being islands in the universe and our feelings of isolation, even of incommunicability. After all, the adventure of Langevin's twins demonstrates that even true twins can, in principle, be changed so that one will no longer be the copy of the other at the present instant. One would be the mirror of the youth of the other, who would in turn presage the age that awaits the first. In another context, Jacques Monod said it in this way: "Man knows at last that he is alone in the immense indifference of the Universe!"[2]

The second lesson of relativity theory corrects the first: all the island-universes remain subject to a common law that imposes common themes on their individual rhythms. These common themes are, therefore, not arbitrary. Propagated causality is this common law. The light cone issuing from an event contains all the events in the world that can be causally linked to it. Reunited, all the light cones weave reality like a network of chain-mail links, giving a structure and a sense to the world.

Einstein's relativity theory no doubt enjoyed much greater success than the timid offerings of Lorentz and Poincaré because of this faith in an order in the world that, in particular, imposes the relative rhythms of clocks. The two Einsteinian principles used in describing nature—that the speed of light is invariable and the frame of reference chosen is irrelevant—are not logically linked other than within this paradigm of causality. This paradigm has become so important in contemporary physics that a physicist who wants to build a theory of a new interaction (between two new elementary particles, for example) always begins by writing that the energies of interaction must obey Lorentz transformations in any change of the system of coordinates. Einstein's realism is thus, above all, a realism of causality. There is no other possible explanation for the extraordinary predictive power of the principles he proposed.[3]

■ Causal Time Is Not Oriented

But the causal interpretation of time strips it of its orientation, of its arrow.

Two particular events in the four-dimensional universe of space-time are or are not causally linked depending on whether one of them is situated at the interior or the exterior of the light cone that can be drawn with the other as its origin. The theory of relativity by itself does not indicate an order between causally linked events; it does not designate events as either causes or effects. Defining time's arrow requires introducing an element alien to the formalism and decreeing that causes are *in principle* anterior to effects. Certain epistemologists decry a pure anthropocentrism in this new principle. According to them, situations in which causes of an event would be situated in the "future" part of its light cone appear absent only because human intelligence cannot detect them. Humans always develop their discourse by starting from causes to deduce effects.

The simplest example of a phenomenon determined by an anticausal solution, which is allowed in principle by the laws of electromagnetism, would be the detection of spherical electromagnetic waves being spontaneously absorbed by an atom. Nonetheless, to interpret such a phenomenon, it would be necessary, through a sort of prescience, to be able to designate its "cause"—the atom that will absorb the waves—even though this cause is manifested only at the end of the procedure and not at the beginning. The stance that science takes in considering only past causes exercising their effects in the future must be noted to be an absolutely crucial choice. It determines in fact the boundary between physics and metaphysics, between materialistic science and religious faith, because religion does not exclude final causes, which appear, on the contrary, as God's "reserved domain."

Despite the abstract nature of this subject, it is important to be more specific about what has just been said of the propagation of light and other electromagnetic waves. In general, these are emitted by a pinpoint source in the form of spherical and concentric waves that diverge on leaving the source. The inverse phenomenon of naturally converging spherical waves "merging" with, or being absorbed into, a material particle is not found in nature. However, Maxwell's equations allow a priori both types of solutions. It seems, thus, arbitrary to impose the supplemental condition—that of causality[4] or of "retarded waves"—which rejects the "advanced waves" as physically impossible.

It is possible, however, that this seemingly arbitrary condition on causality in electromagnetic theory provides a shortcut, a way of describing electromagnetic waves that takes into account the conditions of their formation in our concrete universe, subjected as they are to the law of cosmological expansion. More precisely, one cannot exclude a priori the possibility that the apparently arbitrary and anthropocentric condition of initial causality is no more than a synthetic fashion of summing up our situation in an asymmetrical world by way of its limiting conditions. The laws of physics would be *de jure* symmetrical in relation to the direction of time's passage or, in other words, unaffected by the arrow of causality, but the "boundary conditions" would not be symmetrical, for the universe is in fact expanding.

In fact, John Wheeler and Richard Feynman showed around 1949 that two equivalent ways of presenting the solutions to Maxwell's equations exist:

1. In the usual presentation, only the retarded solutions are kept, and the advanced solutions are rejected as not obtaining in the physical world.

2. *On condition that the universe is opaque in the future,* the advanced solutions and the delayed solutions are added, each with the proportion 1:2. In this case, it is the reaction of the absorber in the future that nullifies the advanced part and gives a weight of "one" to the delayed solution.

The universe's future opacity would require that all waves emitted must one day be absorbed. This question of opacity in the future and in the past, according to which any "advanced" wave emitted today points forward or backward in time to a situation of opacity in the universe, seems thus crucially linked to the problem of causality.[5]

The universe is manifestly opaque toward the past as long as the Big Bang model is valid, or at least as long as the interpretation of background radiation is correct. This background radiation testifies to the existence of a past epoch in which matter and radiation were in equilibrium.

But it is not known whether the universe is equally opaque toward the future. A photon emitted today may very well traverse cosmic distances without being absorbed by any atom of galactic or extragalactic matter. Indeed—if the expansion of the universe is rapid enough and the density of matter small enough—it may *never* be. The universe would then be transparent toward the future. Sadly, this uncertainty considerably weakens the conclusions that can be drawn from the reflections of Wheeler and Feynman.

In 1973 an attempt was made in the United States to bring new elements into this debate using a curious experiment. The Wheeler-Feynman theory was used to try to determine whether the universe is transparent or opaque toward the future. The experiment, conducted by the American physicist R. B. Partridge, consisted of emitting radio waves toward space using a directional antenna and comparing very precisely the device's electrical consumption when the antenna was directed toward an apparently "empty" zone of space (apparently nonabsorbing, but let us not forget that the space we see reflects its past state, when the active experiment was conducted to interrogate its coming state) to the electrical consumption when the antenna was covered with as absorbent a screen as possible. According to the Wheeler-Feynman theory, if the universe is opaque toward the future, the advanced waves emitted must be totally extinguished and the retarded waves brought to their maximum intensity in such a way that the electrical consumption would be the same in the two cases; if, on the contrary, the universe is transparent toward the future, the emitter's electrical consumption should be lower when aimed at open space.

They therefore examined with particular care the power drawn by the transmitter. No difference was registered up to a billionth of a measure. The conclusion of the experimenter was that, unless the Wheeler-Feynman theory is incorrect, absorption along the light cone toward the future was complete to better than one part in 100 million.[6] Certain theoreticians, however, dispute this conclusion.[7]

We do not yet know whether it is possible, using the Wheeler-Feynman theory or its eventual extensions, to account for the absence of advanced waves and for our existence in a world where physical phenomena are completely dominated by initial causes. If matter-radiation interactions take place in a symmetrical way at the two ends of the chain of time, the asymmetrical propagation of causal electromagnetic waves (retarded waves) or anticausal waves (advanced waves) should not be possible. A still unknown mechanism that could disrupt this symmetry would then have to exist. Indeed, in the actual state of the theory, it seems that only advanced waves would be allowed to propagate—to the extent that the only unsymmetrical solution compatible with our knowledge of astrophysics is that of a universe opaque toward the past and transparent toward the future—while the propagation of retarded waves would be prohibited. This is obviously contrary to the most elementary daily experience. We are approaching here the speculative frontiers of current physical theory.

■ **Quantum Inseparability**

The second difficulty with the causal interpretation of time is linked to the development of quantum mechanics. This difficulty was raised for the first time by Einstein during his reflections on the principle of causality and its relationship with the principle of reality. He thought at that time that the theory of quantum mechanics as it had been developed, in particular by Niels Bohr and the Copenhagen school, was necessarily incomplete. When rereading the article published by Einstein, Boris Podolsky, and Nathan Rosen,[8] most present-day physicists conclude differently. They underline the inadequacy of the criteria of reality that the authors proposed and, more precisely, the weakness of the authors' relativistic principle of causality, which supposes all properties of the objects of physics to be local. In fact, the response that we would give today to the question asked by Einstein and his colleagues is of a larger philosophical significance than even they suspected. It would be stated thus: Quantum mechanics is perhaps a complete theory. But physical reality does not have the property of locality. It cannot be separated into individual objects each assigned its own objective individual characteristics. It is inseparable.

Einstein's argument is a strong one. It marked the end of a long maturation whose traces can be found in an oral presentation at the Fifth Congress of Solvay in 1927 and whose refinement can be followed throughout the discussions he had with Niels Bohr during the years 1930–33.

In 1927 Max Born, who invented the probabilistic interpretation of quantum mechanics, showed that each particle must be associated with a wave, called a quantum wave. The fundamental equation of quantum mechanics (the Schrödinger wave equation) describes the behavior of this wave and not the behavior of the particle itself. The wave is not material: it indicates only the probability that the studied particle will appear at a given location if its location is sought at a given instant.

Commenting on Born's exposition, Einstein proposed the following thought experiment. Imagine the motion of a particle that is moving from left to right and that is made to traverse a small diaphragm placed perpendicularly across its course. Imagine also that we have, at some distance to the right of the diaphragm, a scintillation screen or a photographic plate. After having passed through the diaphragm, the particle will make its mark on this detector. Calculations indicate that the probability wave representing the particle as it exits the diaphragm is a divergent spherical wave propagating in all the space at the right of the diaphragm (the phenomenon of diffraction). According to Born, the intensity of this wave at a given point on the screen of materialization represents the probability of observing the particle at that point. Let us choose to give the detector screen a hemispherical form centered on the opening of the diaphragm, like a bowl for gathering the diffracted particle. In this case, the wave of probability would simultaneously arrive at all points on the screen. How then can it happen that the particle always materializes only at one point and never at two, ten, or twenty points? Must we envision an instantaneous influence, one more rapid than the speed of light, that "warns" the other possible points of impact that the particle has materialized in such a spot and must not, therefore, appear elsewhere?

In his argument Einstein admitted in fact that the particle is necessarily somewhere throughout its trajectory between the diaphragm and the screen, although quantum mechanics and its "probability wave" are incapable of pinpointing it. Quantum physicists do not speak of the "position" of the particle except at the points at which it is not only characterized (for example, just at the exit of the diaphragm) but truly "measured," that is to say, observed concretely. Disagreeing with them, Einstein reclaimed the right to speak of the particle's "position" at every instant of its existence, even if the experimental means to determine that position do not exist.

However, Einstein's thought experiment is complicated by its inability to describe minutely the mechanism through which the particle materializes at its final position. The point of impact results from a photochemical process so complex that one can sincerely dispute the contention that the appearance of a luminous point on the scintillator or of a speck of silver on the photographic plate truly translates the property of "position" that one would be obliged to attribute to the particle even if the detector were not placed along its trajectory. It is well known that the process of measurement can disturb the object of the observation. It cannot be absolutely affirmed that the screen plays only a passive role and that the position detected would be the same in the absence of the screen. This is why the resolution of the debate had to await the development of more sophisticated experiments.

As long as the notion of an inseparable reality remained theoretical and confined to thought experiments, it did not truly bother physicists. The situation changed in the 1970s, however, when it became physically possible to test quantum inseparability.

■ Experimental Verification of Inseparability

At the Orsay Institute of Optics, at the beginning of the 1980s, Alain Aspect conducted a series of experiments designed to prove quantum inseparability and to study some of its properties.[9] These experiments demonstrated that the states of polarization of two photons emitted during certain atomic processes cannot be attributed to each photon separately.

An analyzer (a polarized filter) was placed along the trajectory of each photon, and inseparability was manifested when these filters were given different orientations. For example, the axis of polarization of the first was vertical and that of the second was inclined by about twenty degrees. Inseparability remained even when the orientation of the analyzers was chosen *after* the emission of the two photons and when the distance between the two analyzers was such that no signal could go from one to the other after the passage of the first photon and before that of the second.

The conclusion of this experiment and others of the same type was that nature manifests a fundamental inseparability, so that two systems having interacted in the past (the two photons emitted by the same atom in Aspect's experiment) are always interdependent, whatever their distance. No complete description of either system is possible without reference to the other.

The properties that manifest the phenomenon of inseparability are those that quantum mechanics describes as "incompatible," which means that they cannot be known simultaneously when measured on a single particle. These are the properties of polarization of the photons measured by differently oriented analyzers or, for material particles, the properties of position and momentum projected on one axis, and more generally, the properties that describe the particle as a wave on the one hand and those that evoke a pinpoint object on the other. The thought experiment proposed by Einstein in 1927 was of this latter type. Current electronic techniques have recently allowed an improved version of it to be realized.

Using an idea from Wheeler, two groups of researchers conducted an experiment where the decision to measure either a pinpoint property (the position of a photon) or a wave property (observations of interferences between two branches of an interferometer) is made after the introduction of the photon into the experimental setting (fig. 9).[10] The photon first encounters a semitransparent mirror whose function is to divide the "beam" into two parts of equal intensity. The two beams follow different paths along branches A or B, but they are finally superimposed on the observation screen at the exit of the apparatus. The intensity observed on the screen O is a function of the difference in path lengths between the two branches of the apparatus, a classical effect of interference.

First note that the interferences are observed even when the intensity of the beams is diminished to such a point that the photons traverse the apparatus one by one. In this case, of course, we can make judgments of interferences only after a count of the impacts on the observation screen O shows that a sufficient number of photons have traversed the apparatus. Each photon thus divides, in a certain way, into two waves propagating in the two separate branches of the interferometer before recombining on the screen. The result of the second part of this experiment is even stranger. On the trajectory of one of the two branches is placed a rapid point switcher (Pockels cell) capable of diverting the photon, if it is present in that branch, and sending it toward a counter P that will confirm its presence on that trajectory. This counter acts exactly as a counter that might have been placed in Einstein's experiment at a precise point somewhere between the diaphragm and the

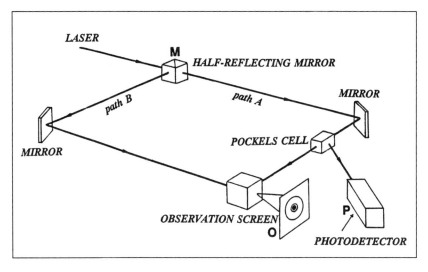

Figure 9. Sketch of the experimental arrangement corresponding to the experiments on the separation of a beam made from a single photon, inspired by an idea of J. A. Wheeler. The quantum wave corresponding to the photon is separated into two branches by a semisilvered mirror *M*. The Pockels cell placed on trajectory *A* can, in a very short time, divert the corresponding half-beam toward a photodetector, where the photon is eventually absorbed and detected. When the cell is not in use, the two half waves recombine into *O*, where the effects of classical interference, a function of the difference in path lengths between the two half-beams, can be seen. When the Pockels cell is in use, the incident photon behaves as a single particle and is detected on average one time out of two on trajectory *A* by the photodetector *P*. The paradoxical aspect of the experiment consists in that these conflicting behaviors are not changed even when the Pockels cell is activated *after* the incident photon has interacted with the mirror *M* (and before the photon has had the time to reach the Pockels cell). From A. Shimony, "The Reality of the Quantum World," *Scientific American* 258 (Jan. 1988): 36. Used by permission of Abner Shimony.

screen. The experiment's refinement consists in that the Pockels cell can be activated after the laser beam has interacted with the semireflective film and therefore after the photon has been "separated" into two beams.

The experiment's results were as follows: if the cell is not activated, the photon always behaves as a wave; there is interference between the two branches. If the cell has been activated, 50 percent of the time a photon's presence is observed in branch A. But if this is the case there is never a photon in branch B; the detector O remains mute. Although there is no possible causal interaction between the photon's materialization in the counter P and its materialization in the counter O, the "decision" of the beam "in branch A" to behave like a particle is instantaneously understood by the beam in branch B as an order to "vanish." The photon thus has non-local properties that concern both branch A and branch B at the same time.[11]

Inseparability is today an established experimental fact. Although the process of seeing it is intricate and necessarily demands that observations be repeated on a large number of systems, it is not a limited property that can be neglected even in a first approximation for distant systems: it has, in fact, been verified for all systems that have interacted in the past, whatever their current "distance."

■ Inseparability: A Radical Property of Reality

Until very recently, the success of relativity was such that it obscured and relegated to second place the objections and reservations of physicists and philosophers of the interdependent tradition, those who suspected, like Spinoza, that the separation we impose on things is fragile and secondary. Spinoza strove to show that this separation was no more than the subjective reflection of our desire to reign over nature, to "appropriate." However, wave mechanics had already considerably reinforced the doubters' position. Not that the waves superimposed on the particles raised doubts about space or separability; almost all physicists admit that the waves of quantum mechanics are "nonphysical," since they are simple algorithms, waves of probability. But quantum mechanics introduced an interdependence between the object studied and the instruments of measurement that did not exist in classical physics. Niels Bohr used to insist on this new aspect of the philosophy of nature. The properties of a quantum object cannot be discussed without specifying the instruments by which it is to be observed.[12]

Einstein, understanding the danger that the new concepts brought to bear on the realism of the separable—that is to say, on the idea that the systems of nature separated in space possess properties that are their own—fought them furiously. But the conclusion drawn today from these experiments differs from his own: it is the hypothesis of separability that must be abandoned, and quantum mechanics has not been proven to be wrong or incomplete. In 1975 Bernard d'Espagnat[13] showed that the results obtained did much more than simply confirm or invalidate the fundamental principles of quantum mechanics. They tend in fact to demonstrate the fundamental inseparability of nature, independent of any particular theory.

More precisely, the experimental results imply that at least one of the following postulates is false:

1. the postulate of *local realism:* the systems possess local and intrinsic properties that are unaffected by the decision to make the corresponding measurement;

2. the postulate of *persistence of intrinsic properties:* the properties ascribed to a system remain true as long as the system remains isolated (except in the case of spontaneous decay);

3. the postulate of *nonretroactivity:* an intrinsic property verified in a given system cannot be changed by modifying the instruments with which it will interact in the future; or

4. the postulate of *hierarchical organization:* an intrinsic property verified in a system is not affected if it is attached not to the system itself but to another larger system containing the first. For example, the intrinsic properties of an electron are not changed if the electron is considered as itself or as one particle in an atom.

At the current time, no one thinks seriously of doubting the fourth postulate, which applies only to purely mental exercises; these do not exclude in any way the possibility that "small systems" can have physical interactions with "large" ones. Opening postulates 2 and 3 to questioning would shake the foundations of even the possibility of a physical theory, introducing into nature a whimsicality or a finality incompatible with the program of science, which is the search for causal laws. Suspicions thus fall naturally upon the first postulate, that of separability.

The nonlocal properties of reality brought to light by quantum mechanics lead us to reexamine both the implications and significance of the principle of causality and the identification of time as a parameter of causality, as proposed by relativity.

■ Separating Causality and Time

Bernard d'Espagnat expressed his conviction on the need to abandon the principle of separability and its consequences in these terms: "Within the framework of a realistic conception, I don't see, on my part, any solution other than to abandon the principle of separability. This signifies, schematically, either that certain systems currently distant from each other must be considered as constituting a single system or that between distant systems there exist influences faster than that of light."[14]

This formulation, however, raises a question. The formulation of the alternatives—total solidarity of systems or influences more rapid than light between two separate systems—are they truly two equivalent ways of expressing the same situation? This is not certain, and the second possibility seems in any case very rash insofar as it can suggest that the laws of relativity are sometimes transgressed. Everyone recognizes, in fact, that the experiments based on Einstein, Podolsky, and Rosen's "paradox" manifest instantaneous *influence* at a distance but in no case allow information-bearing *signals* to be transmitted at a speed superior to that of light. This results from the strictly aleatory character of the observations made in any one place. For example, an observer close to one of the two analyzers in Aspect's experiment never registers other than aleatory responses conforming to the laws of probability, whatever the orientation of the analyzer examining the state of polarization of the photon in the apparatus's other arm. It is only when using the statistics concerning measurements effected on the two photons at once that the characteristic correlations of inseparability can be seen. Thus, the "influence" between the orientation given to one of the analyzers and the responses obtained at the other (and reciprocally) does not constitute a "signal." In a context known to be characterized by nonseparability, it thus seems better to avoid speaking of transmission of influences between "distant" objects, when the notion of distancing presupposes that of distance and therefore that of separability.

How does the fundamental inseparability that henceforth will characterize contemporary physical theory affect the status of causality? Does nonseparability bring determinism back into question? Physical theory requires determinism, and the principle of relativistic causality is still fully valid when applied to the operational level of phenomena. But, as underlined by Bernard d'Espagnat, the principle of relativistic causality can no longer, in any way, be understood as being attached to reality as such, which eludes the categories of space and time founded on relativity. In the same way, it can no longer establish an ontology of time. The ideal of relativity, reducing time to the expression of a law of causality propagated in nature in itself, must be abandoned. Time must be separated from causality.

In quantum mechanics, the identification of time as a parameter of immediate causality (in Schrödinger's equation, for example) poses a problem that has long been recognized. Time had been "imported" unchanged into this theory from classical mechanics as a simple parameter and not as an operator, like other measurable physical quantities. Several years ago David Bohm proposed introducing a parameter of causality distinct from time into quantum theory.[15] More recently, Ilya Prigogine defined quantum time, a quantum operator linked to time-becoming.[16]

This disjunction between time and relativistic causality also opens the door to different approaches. Olivier Costa de Beauregard remarked that in quantum mechanics the continuity of displacements in space-time has been broken, since all phenomena are analyzed in terms of "quantum jumps," and that, therefore, there is no reason to exclude the possibility of abrupt jumps from an "orthochrone" space-time, in which time flows in the direction we habitually give it, to a space-time in which time flows in the reverse direction.[17] This position makes very subtle distinctions, for it is difficult to bear in mind, when speaking of "time" without any further explication, which concept of time is meant—the primitive notion or an expanded relativistic concept. We might as well seek new, completely atemporal expressions of the principle of causality, even if it entails abandoning the direct link between causality and time. Some attempts have been made in this direction by trying to express causality within the framework of conditional probabilities.[18] In this view, causality indicates only that the probabilities of realizing an event are not the same, depending upon which conditions are fulfilled, without temporal priority. These attempts, in which Costa de Beauregard is participating, are still too new or too fragmentary for us to judge them.

■ ■ ■

The separation of causality and time made necessary by quantum inseparability raises the possibility of rethinking the relationship between the primitive notion of time and the concepts derived from it in the physical and natural sciences. To sustain this reflection, we must turn once again toward the facts, scrutinizing them with all the skill and delicacy our modern means of investigation provide.

Part 3

From the Atom to the Universe: The Emergence of Becoming

The Suspended Time
of the Atoms

Up to this point I have looked at the construction of a concept of physical time from a historical perspective. I have traced the development of a causal interpretation of time, with its difficulties and limitations. Now, taking another approach, I will examine the implications of physical time by climbing the ladder of the dimensions of systems, from the smallest to the largest, from one extreme to the other. We will see that time does not apply to elementary particles, which are ruled by the laws of quantum mechanics, as long as they are considered as closed and isolated systems. For atoms, time is in some way "suspended," to use Lamartine's words. Despite its potential for decay, an isolated atom can teach us nothing about the passing of time.

Radioactive atoms or unstable elementary particles decay spontaneously. The time that passes before the decay of an atom or a particle is extremely variable, but atoms and particles of a determined type have an average lifetime characteristic of that type. Thus, after a period of 6.5 billion years, half of the atoms of uranium 238 present at the beginning of that period will have decayed. Natural radioactive substance are not abundant, because those whose average lifespans are greatly less than the age of the Earth have, for the most part, disappeared from the surface of the globe. Physicists, however, know how to re-create, through nuclear interactions in laboratories, atoms and elementary particles with an average lifespan infinitesimally shorter than that of uranium. Among these is the pi meson, whose average lifespan in a free form is less than 0.03 microseconds, and the meson $K°$, of which there exist two versions with different lifespans, each on the order of a billionth of a second.

This decay, whether of natural or artificially produced particles, covers a vast scale of time; the very phenomenon of the instability of matter is enough to demonstrate that becoming is inscribed in the heart of things.

Nuclear decay offers an ideal ground to researchers who want to track physical duration, because of the basic nature and simplicity of the quantum-mechanical description of the physical objects concerned.

■ Spontaneous Atomic Decay Is Subject to Chance

When unstable particles decay, they seem to obey a simple descriptive law characteristic of the increase of entropy. This is the law of exponential decrease, whose discovery in the case of radioactive atoms dates from the beginning of the twentieth century. At that time, Marie Curie had already discovered one of the first-known radioactive substances, polonium, and had purified another (radium) from it by patiently concentrating tons of pitchblend in the shed of the Ecole de Physique et Chimie Industrielles (School of Industrial Physics and Chemistry) of Paris. In 1903 the activity of uranium, thorium, and radium and of the products of their decay (the gaseous "emanations" of thorium and radium, as they were called at the time) were subjected to systematic measurements in various laboratories, among them that of Ernest Rutherford and Frederick Soddy in Montreal. The activity of these substances was measured by placing a sample between the plates of an electrical condenser. The ionization induced by radioactivity caused the air or other gas between the capacitor's plates to conduct a charge. The discharge current was proportional to the number of ions created by the rays between the two plates.

During their observations, Rutherford and Soddy noted that each time they used a pure radioactive substance carefully separated from the other radioactive substances that gave it birth or to which it gave birth in its turn, the activity measured (the current of ionization) always decreased exponentially with time:

$$I(t) = I_o e^{-\lambda t},$$

meaning that the number of radioactive atoms present in the sample decreased exponentially with time as long as each decay produced the same quantity of ions. It could also be said, by mathematically deriving the above equation with respect to time, that the proportion of radioactive matter which transforms itself during one unit of time is a constant. No outside influence other than the number of atoms present at the instant t determines the number of atoms that will transform themselves in the successive unit of time. Only the quantity of radioactive substance present appeared able to change the number of transformations observed. Rutherford and Soddy concluded that radioactivity was a specific property of the element considered, one that gave rise to its own destruction. The constant λ, which characterizes the speed of this spontaneous "decay," is called the rate of decay for the substance studied.

Rutherford and Soddy did not comment further upon the material's astonishing propensity for self-destruction, which they had just proved and which reawakened the question of matter's stability and its immutability over time. However, in 1903 Einstein had not yet proposed replacing the law of the conservation of matter with that of energy.[1]

Following the work of Rutherford and Soddy, the exponential law of radioactive decay was almost always accepted without reservation. Usually, an exponential decay characterizes phenomena of relaxation, corresponding to the sudden discharge of slowly accumulated energy, which is dissipated in the form of heat. These phenomena are frequent in mechanics and electronics. When, at any instant, the flux of discharge (e.g., electrical current) is proportional to the potential (e.g., difference of potential), then the current, like the potential, decreases exponentially. The mathematical proof is simple. The phenomenon can even reproduce itself periodically through a repetitive succession of charges and discharges. This is what could be called a "thermodynamic clock," in contrast to "mechanical clocks" based on the phenomena of sustained oscillations. Thermodynamic clocks work by dissipating energy; by contrast, mechanical clocks require that the phenomena of dissipation (the mechanical friction or the electrical resistance) be reduced to a minimum. In practice, the exponential decrease law is the signature of a phenomenon of relaxation. Inversely, any phenomenon to which is attributed the phenomenon of relaxation should obey a law of exponential variation in time.

The decay of unstable particles produces waste heat. At the same time, it leads to the appearance of at least two types of atoms at the heart of the sample, since some are transmuted and others are not. The detector gives the number of atoms that have decayed, but it does not indicate precisely which ones these are. Evaluated in terms of information, the corresponding ambiguity allows estimations of the minimum increase in the corresponding entropy:[2] the radioactive sample indeed constitutes a thermodynamic clock.

The decay of unstable particles, however, does not seem to be a phenomenon of ordinary relaxation for two reasons. First, flux and potential must be proportional in relaxation, but it is difficult to see here why the flux (the number of decays per second) should be proportional to the potential (the mass of the sample). Second, the theory has difficulty justifying the exponential decrease law and tends, rather, to predict a slightly different law.

In order that the number of decays per second be proportional to the mass of the sample observed, on the one hand *all* the atoms of a radioactive sample existing at the instant t must have the same probability of decaying during the next time interval; on the other hand *each* atom, taken individually, as long as it exists in a nondecayed state at the instant t, must have the same probability of decaying during the next time interval, *whatever the instant considered*. These two conditions must not be confused with each other. The second is the more interesting from our point of view, for it denies that the concept of time or duration has a meaning when applied to a single atom. It expresses, in fact, the irrelevance to each atom of its own past. It affirms that duration is a collective phenomenon, a phenomenon that necessarily deals with groups: briefly, it is a thermodynamic phenomenon.

This condition, known as *Fermi's golden rule,* was demonstrated within the framework of quantum mechanics. But the demonstration in question uses an approximation that cannot be completely justified. As a consequence, radioactive decay might not rigorously obey the exponential decrease law.

The demonstration of Fermi's golden rule assumes that the total energy of the products of decay is able to take any value, positive or negative, although that energy is of course always positive. In principle, however, the approximation made seems of little importance. The law of the conservation of energy ensures that the final energy of decayed states is always very close to the original mass (positive) of the atom or particle before its disintegration. The eventual difference is limited by Heisenberg's fourth uncertainty principle, which links the possible differences in the energy balance sheet to the lifetime of the atom or unstable particle. If the strictly positive boundary condition for the final energy is taken into account in the calculation, the probability of decay is not constant, as Fermi's golden rule would have it, but rather decreases slightly over the course of time.[3]

Thus, over very long periods of time the observed decay rate should be somewhat less than that predicted by the purely exponential law. The atom would, so to speak, "age" slightly and would have less and less chance of decaying. But the difference in what was foreseen by a purely exponential law would not be perceptible except for atoms with actual lifespans on the order of thirty or more times the average.

In fact, the difference in behavior is so minimal that it has not yet been observed experimentally. Imagine an experiment using a particle accelerator to test the exponential law of decline with all the required precision— for example, by observing the decay of a billion pi mesons. It would be a difficult experiment, requiring several years of work. Would the demonstration of such a small difference in a simple law merit such an effort? Certainly, for the epistemological importance of such an experiment is immense. If a deviation from the exponential law of decline were observed, it would mean that the "elementary" particles are not truly indifferent to the passage of time; the affirmation that duration is a paradigm valid only for large groups of atoms, a macroscopic concept, would be disproved. If, on the other hand, the measurements were to confirm the accuracy of the exponential law of decline despite the corrections predicted by the theory, the collective character of duration would be confirmed but would lead inevitably to the conclusion that the variables "energy" and "time," described as paired in quantum mechanics, are not the variables our instruments actually measure. It would therefore be possible to redefine them, for example, by envisaging a "quantum time" distinct from classical time, to which it would be linked by a sort of principle of correspondence. Ilya Prigogine and his group have already developed an approach of this type.[4]

■ Reversing Time: A Theoretical Operation

The time of quantum mechanics and classical time must be distinguished from each other when it comes to the *reversal of time*.

The terminology introduced here by theorists of quantum mechanics is certainly not very fortunate, for the expression "reversal of time" is contrary to the very definition of the primitive notion of time. Even in relativistic physics, which envisions situations, environments, and physical experiments that demonstrate a significant change in a clock's rate of operation, no situations are encountered where the direction of time variation is reversed. After his voyage, Langevin's twin could live on the Earth of the twenty-second century but would never be able to live again at the time of François the First. Relativistic time moves always ahead, even if all the clocks do not keep to the same second.

The expression must, therefore, be taken in a metaphorical sense and not literally. Reversing time means imagining an extraordinary creature with an "inverted" memory remembering the future[5] and asking whether, for that being, the laws of physics would differ from those we know. In practice this comes back to changing time, velocity, and currents to their opposite values whenever they appear in the fundamental equations of the theory. If the laws remain the same after these transformations, it could be concluded that the past and the future are interchangeable from the point of view of physical theory: we could speak of a symmetry of nature through reversal of time. If they differ, we could say that this symmetry is absent or that it is "violated." An absence of symmetry through reversal of time implies, of course, that nature is not indifferent to the direction of time's passage and that the "arrow of time" thus has an objective significance.

In fact, all the basic laws of physics obey symmetry by time reversal, whether in classical mechanics, quantum mechanics, or relativity. Consequently, the asymmetry observed in the evolution of macroscopic material systems whose study is the object of thermodynamics must be due to very specific boundary conditions, which are incompletely explained to this day, linked perhaps, as will be seen in the next chapter, to the current phase of expansion of the universe.

Furthermore, in the 1950s Julian Schwinger, Gerhard Lüders, and Wolfgang Pauli set forth a theorem showing, based on very general hypotheses, that there exists in nature a global symmetry called "CPT symmetry." This symmetry means that applying the three following operations to whatever system in whatever order would result in a system that, from the point of view of physics, could not be distinguished from the first:

1. Substitution of antiparticles for particles (the protons are replaced by antiprotons, the electrons by positrons, the π^+ mesons by π^- mesons, etc.). This operation is called "charge conjugation" and is usually represented by the letter C.

2. Right/left inversion, as if the scene were viewed in a mirror. This is the operation called "parity." It is usually represented by the letter P.

3. Reversal of time, that is, the scene is imagined as viewed by a witness with inverse memory. This operation is usually represented by the letter T.

To make these abstractions more concrete, let us see how, for example, it might be equivalent to speak of ordinary particles reversing their way through time or of antiparticles following time in the usual sense (Feynman). Suppose for an instant that in nature there exists an uncoupling between the temporal order of succession, which we will call t, and the order of causality, which we will call Θ. Figure 10 represents the "trajectory" of a particle, whatever it may be.

In the figure on the left, this trajectory follows its natural path according to the directional order of causality and the directional order of temporal succession. The simple image is that of an ordinary particle existing without accident from the instant t_1 to the instant t_8.

In the figure on the right, the order of causality is no longer identical to the temporal order. The trajectory represents the path of an ordinary particle "going backward in time" between the points $\Theta = 5$ and $\Theta = 7$ of the parameter of causality. It can also be interpreted by respecting the ordinary passage of time; at the point C, $\Theta = 7$ ($t = 3$), one can observe the creation of a pair consisting of a particle (on the right-hand branch) and an antiparticle (on the left-hand branch). This last annihilates itself at the time $t = 5$ with the ordinary particle whose trajectory is represented by the segment AB. Thus, this figure, which represents for a real observer a complex phenomenon of creation and annihilation bringing into play two particles and one antiparticle, could also be interpreted by an observer freed from time's arrow as the trajectory of a single particle zigzagging in time.

If the CPT theorem is valid, a phenomenon of nature asymmetrical in one or the other of the three operations C, P, or T cannot be symmetrical for the

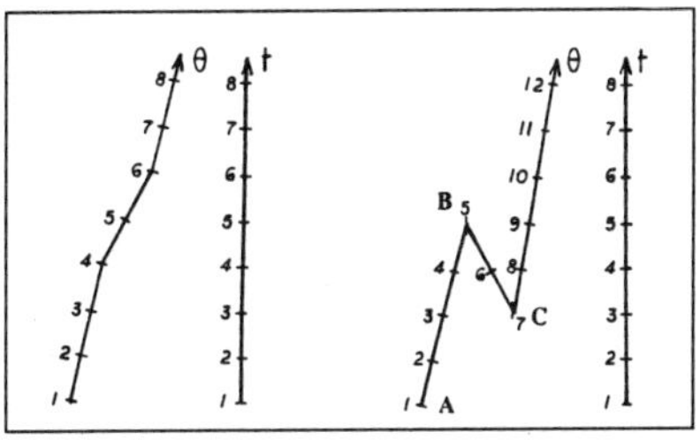

Figure 10. A possible nonparallelism in microscopic physics between the time t and the parameter of causality Θ can lead to the observation of the creation and annihilation of antiparticles.

remaining two operations, PT, CT, or CP. In 1964, however, an exemplary phenomenon was discovered in which there was manifestly no conservation of the CP symmetry. It concerned the decay of the long-lived K° particles. According to the CPT theorem, then, there should be no type T symmetry either, that is, no symmetry with respect to time reversal. Even though the phenomenon concerns only certain very precise types of interactions between elementary particles, it seems to say that the arrow of time exists not only in thermodynamics but also at the level of individual particles.

■ A Case of an Asymmetrical Phenomenon: The Decay of the K° Particles

The K° particles were discovered in 1951. Created in nuclear collisions as cosmic rays enter the atmosphere, they have a lifetime of only about ten billionths of a second (for the variety called "long-lived"). Since the development of particle accelerators, the properties of K° mesons have been abundantly studied. They decay by weak interaction, most often into three lighter particles (among the pi mesons, the mu mesons, the electrons, and the neutrinos). Since 1956 it has been known that weak interactions do not obey either the symmetry by parity (operation P) or the symmetry by charge conjugation (operation C), but until 1964 it was thought that they responded to the joint CP symmetry and thus also to symmetry by time reversal. This was, it was thought, the reason that the long-lived K° meson did not decay into two pi mesons, for the theory shows that in this case CP symmetry would not be respected.

That same year, in the American laboratory of Brookhaven, J. H. Christenson, J. W. Cronin, V. L. Fitch and R. Turley, who studied the interactions of the long-lived K° with hydrogen, observed that a small fraction of the long-lived K° mesons decayed into two pi mesons. The weak interaction responsible for their decay did not, therefore, perfectly respect CP symmetry.

If the CP type symmetry is not respected, even in a small degree (the extent of violation in the decay of the K° mesons is only a few percentage points), and if the CPT theorem is valid, then there must also exist, in the decay of the K°, a mechanism violating symmetry by time reversal.

Thus, processes such as the decay of long-lived K° mesons into two pis lead to the following dilemma. Either these processes are not reversible, not indifferent in terms of the flow of time, in which case nature is furnishing us, at the elemental scale and independently of all subjective reference, an absolute standard for the direction of time's flow, although the phenomenon in question very likely has nothing to do with the aging of matter acknowledged by the second law of thermodynamics; or the CPT theorem is not valid because one or the other of its basic hypotheses is not fulfilled in nature. One of these hypotheses, perhaps the least sound, is that the field equation ruling the evolution of a quantum system must be local, that is, it must be possible to describe the behavior of the system solely from the basis of the values of the quantum fields at the very point where the system is located. Do not recent discoveries on the nonlocality of quantum mechanics (see chapter 11) open the door to a possible disproof of the CPT theorem? Theorists will have to address this question.

■ Direct Tests of Temporal Symmetry

To shed some light on this dilemma, other tests have been envisioned that might more directly prove a possible violation of temporal symmetry in particle physics. The most accessible test is certainly to measure the neutron's "dipolar electrical moment." This particle is totally neutral in terms of its electrical charge, but it could contain positive and negative electrical charges in an equal number, one slightly set off from the other. If the centers of gravity of the positive and negative charges did not coincide, the neutron would have a nonnull dipolar electrical moment, but then symmetry through reversal of time would not be respected.

The neutron is an elementary system endowed with symmetry of revolution. Its intrinsic angular moment, or "spin," defines one and only one preferential direction in space. The vector between the centers of gravity of the positive and negative electrical charges, if it existed, would align itself along this axis. Placed in an electrical field, the dipole would have a tendency to align itself in the direction of the field, exerting a slight torque on the system. Given these conditions, if the direction of the external electrical field were not to coincide with the direction of the spin, the system would effect a precession around the direction of the electrical field, like a spinning top subjected to a slight pull.

Faced with this scene, what would our witness with the inverted memory observe? Such an individual would see the precession take place in reverse, whereas the electrical field produced by a fixed charge placed at a distance would not have changed. The same electrical field and the same electrical dipole would thus determine opposite effects for a normal observer and for an observer with inverse memory. If the neutron were endowed with a dipolar electrical moment, there would be no symmetry through time reversal. It should be noted that at the inverse of the electrical field created by a fixed charge, the magnetic fields created by currents are inverted for an observer endowed with an inverted memory, so that nothing bars the existence of magnetic moments of elementary particles.

Several teams throughout the world have conducted experiments attempting to detect, with extreme precision, a possible dipolar electrical moment of the neutron. Because of their magnetic moments, neutrons placed in a magnetic field exhibit a precession whose frequency can be measured. It is, thus, a question of detecting the slight variation in the frequency of precession that would be induced by their dipolar electrical moments when an electrical field is added to the previous magnetic field.

The most recent measurements give a value less than 10^{-24} e.cm as an upper limit of the dipolar electrical moment of the neutron. Such a value appears to rule out a violation of symmetry through time reversal in the case of the neutron, for it is thought that the quarks that might form this particle should have dimensions on the order of 10^{-15} cm and a charge equal in absolute value to $\frac{1}{3}$ or $\frac{2}{3}$ of the electron's.

Another direct test of symmetry through reversal of time consists in searching for isolated magnetic charges. These are called "magnetic monopoles." The possibility of their existence was suggested as early as 1930 by P. A. Dirac to restore symmetry to Maxwell's equations, which introduce elementary electrical charges but not elementary magnetic charges. The difficulty and the interest lie in the fact that magnetic monopoles, like electrical dipoles, would violate symmetry by time reversal.

Consider, for example, a magnetic monopole placed in front of a loop of current and its associated magnetic field. If the magnetic field, and thus the current in the loop, is of the right direction, it will attract the monopole, which will exhibit a uniformly accelerated motion along the direction of the magnetic field. For an observer with inverse memory, the current, and thus the magnetic field, is inverted, but the monopole still exhibits uniformly accelerated motion in the same direction as before, although the initial velocity (if any) is inverted. The direction of the magnetic field and the acceleration of the monopole would thus be in opposite directions, in contrast with what would be seen by a normal observer. Symmetry through time reversal would not be respected.

Magnetic monopoles have been sought at the bottoms of the seas, in lunar rocks, in cosmic rays, and so on. They have not been found, despite a troublesome observation on February 14, 1982, that has for some time been interpreted as the signature of a cosmic monopole passing through a solenoid ring. That day, at the physics department of Stanford University in the United States, a magnetic monopole trap consisting of a superconductive solenoid ring through which a constant current was run registered an abrupt variation in the current's intensity corresponding to the value expected from a magnetic monopole of charge 137 $ec/2$ (the value of the magnetic charge of monopoles as predicted by the theory) crossing through the magnetic ring. The current being constant in a superconducting ring, aside from minor possible losses, the change in intensity without exterior magnetic disturbances can be interpreted only as the result of a magnetic phenomenon: a monopole crossing the ring or an unexplained accident. This unique observation has not been reproduced, however, and it should have been if magnetic monopoles are numerous enough to give some credibility to the observation. It is true that, even if they exist, magnetic monopoles could be extremely rare. Cosmological arguments suggest that perhaps fewer than one monopole per square meter crosses the Earth's path every hundred thousand years.

■ ■ ■

The interest of researchers in demonstrating phenomena that do not respect symmetry through time reversal has been revived, in these last years, by the growing precision of so-called grand unification models, which attempt to unify the fields of nuclear force (strong interactions), electromagnetic force, and weak force (responsible for the decay of unstable particles) at a very high energy level. According to these models, there must be a fundamental identity of nature and a continuity of properties between these different forces of nature. Since we now know that the weak interactions do not obey symmetry for CP operations, and that this has recently been demonstrated also for electromagnetic interactions, it must be the same for the strong interactions. By reason of the identity postulated between all nuclear interactions and the general transgression of the CP symmetry linked to it, this model also predicts that the proton itself must be unstable, although its lifespan could be very long: on the order of or greater than 10^{32} years! Several experiments have recently attempted to prove the natural instability of the proton by keeping watch on several cubic meters of hydrogenated matter. These experiments have been conducted at the bottoms of very deep gold mines or in tunnels that run through high mountains to protect the detector from any influence by cosmic rays.

Like the long-lived $K°$ meson's decay into two pi mesons, the instability of the proton would constitute new indirect evidence of the violation of symmetry through time reversal; so far, however, none of these attempts has brought proof of the instability in question.

Perhaps nature is truly indifferent to the direction in which time flows, and the CPT theorem is not strictly valid. Perhaps the arrow of time is not inscribed at the heart of the laws governing nuclear forces. If this is the case, the real world, including its cosmological expansion, would be just as intelligible, just as comprehensible, for an observer with an inverted memory. We must not forget that the relativity theory does admit the possibility symmetrical to expansion—namely, contraction, followed by the final collapse of the universe on itself and, ultimately, a new singularity, the "Big Crunch." The principle of economy suggests that we not lend to nature more than is indispensable and that we not abandon—without imperative reasons—the comfortable philosophical position that "becoming" was breathed into matter at the initial instant to give it its vitality rather than its having belonged to matter by right. In the sacred texts, is humanity not born of the breath of the spirit on the clay?

13

The Age of Things, Order and Chaos, Entropy and Information

For the ancient savants, the heavens were a place of perfection, harmony, and incorruptibility. The surface of the Earth, on the other hand, manifested evidence of change and duration. A man could experience lived duration by doing no more than closing his eyes and asking himself questions about the passage of time. When he opened his eyes, daily experience taught him that things around him also aged.

Curiously, until recently the development of science tended to reverse this perspective by presenting the panorama of history in the skies while banishing time from terrestrial physics. The cosmology founded on the general theory of relativity and on astronomical observations suggests an unstable universe that exists in a constant state of change and whose current state is inseparable from its history. Conversely, as it developed, experimental physics progressively stripped from time one of its essential qualities, becoming. Finally, the causal interpretation of time tied to the deployment of relativistic space-time (in which each dimension of space and time is basically equivalent in nature) deprived time of its "arrow."

Expelled from physical theories, temporal irreversibility made a discreet return, dating from the middle of the last century, through the intermediary of thermodynamics, or the science of thermal engines.

In his *Réflexions sur la puissance motrice du feu* (Reflections on the Motive Power of Fire), Sadi Carnot explains that he came to see that the transformation of heat into mechanical energy is limited, in practice, by the irreversible direction of heat transfers between bodies at different temperatures: spontaneous heat transfers take place always and exclusively from hot bodies toward cold bodies. This property is not only the archetype but most likely even the basis of temporal irreversibility. *Entropy* is the physical quantity that allows this irreversibility to be quantified in relation to heat transfers.

■ Entropy and Heat

Fourier's equation describing the propagation of heat in continuous media (1811) contrasts with the fundamental equations of mechanics in that it is not reversible with respect to time.[1] But this detail, evident today, was not noticed at first. Other equations describing irreversible phenomena (Fick's equations for the diffusion of matter) take a form similar to Fourier's equation, doubtless because they all, finally, interpret the dissipation of energy in the form of heat. Even the decay of radioactive substances leads at last to a dissipation of heat, as attested to by terrestrial heat, which maintains itself through the decay of radioactive metals in the globe's deepest strata.

Thus the irreversibility of heat transfer came to be considered as the archetypal physical manifestation of temporal irreversibility. This is important because, if it is true, if all the examples of irreversibility seen in physics at our scale[2] can actually be linked directly to exchanges of heat, it must be possible to analyze these irreversible processes in a purely objective way based on thermodynamic values such as heat, temperature, and so on. These values may seem less objective than the other measures used in physics, like mass and energy, since they apply not to individual particles but only to vast groups of particles. In statistical physics, however, heat and temperature do allow a reductionist type of interpretation in terms of the local average agitation energy of particles constituting the body, which is enough to confer on them the necessary "objectivity." The need to define thermodynamic values objectively constitutes a principle of economy that not all epistemologists will accept but that is convenient to use, absent any experimental findings that might oblige us to abandon it.[3]

The irreversibility of Fourier's equation with respect to time does not mean that the laws of nature by themselves make a straightforward distinction between the past and the future. It means simply that systems that would obey a time-reversed Fourier's equation do not exist in nature. In those systems that can be observed, the spontaneous transfers of heat always take place from hot bodies toward cold bodies. More generally, they are always observed to "age" and never to "become younger."

■ The Invention of Entropy

To understand the significance of the concept of entropy, which was developed by the German physicist Rudolf Clausius between 1850 (the introduction of the concept) and 1865 (the introduction of the corresponding term), it is helpful to understand the spirit of the times.

The preceding years had seen work designed to clarify the general law of the conservation of energy in its diverse transformations (mechanical energy into heat, electrical energy, chemical energy, etc.). This work had mobilized a large part of the international scientific community (Sadi Carnot, Robert Mayer, James Joule, Hermann von Helmoltz, etc.) and was not truly finished until shortly before that time. The most general articulation of the law of the conservation of energy must surely be attributed to Helmoltz in 1847.

In 1850 Clausius vigorously credited his compatriot Mayer with having fathered the law of the conservation of mechanical energy and heat.[4] He was obsessed with the universal importance of this new concept[5] and fascinated by the conservation of energy through its various manifestations and transformations. But if energy is the central concept of physical theory, and if it conserves itself, then there is something essential in nature that escapes becoming. Even more than the other laws of conservation, the law of the conservation of energy is written in an atemporal form. In fact, it can be shown that this law is intimately tied to the invariance through temporal displacement of the laws of mechanics: energy is conserved if the equation of motion conserves the same form through any change in the temporal reference point. *Energy* is, par excellence, the symbol of that which lasts and does not change.

Faced with this magical word, Clausius, inspired by Carnot's principle showing that energy not only conserves but also transforms itself and that this transformation takes place in a single direction, coined the word *entropy* in 1865, intending it to be as close as possible to the term *energy*. He considered the two qualities to be so similar in physical significance that giving them similar names was appropriate and even useful. The Greek root of the word means "transformation," but that is only a short form, and it would be better perhaps to translate it by using a longer form. It is not a betrayal of the memory of Clausius to lend to the word "entropy" the following definition: *that which truly changes when everything is apparently returning to the same form.* For a steam engine like the machine considered by Carnot, the machine is said to have effected a cycle when the volume, pressure, and temperature of the steam contained in the engine's cylinder are restored to their initial values. In appearance, everything has returned to its previous state. During the cycle, however, heat is passed from the hot source (the firebox) to the cold source (the condenser). The entropy of the whole system (the machine and its sources) has increased.

Again, the growth of entropy, as defined by Clausius,[6] may seem to be a specific example of irreversibility tied only to thermal transformations, whereas the law of the conservation of energy is general. This is not the case, however. As stated, all the observed examples of irreversibility, at least in macroscopic physics, seem to be directly linked to exchanges of heat.

Clausius's law, therefore, does have the same degree of generality and epistemological importance as Helmoltz's.

The law of increasing entropy was first articulated within the restricted framework of isolated systems, but it has since been generalized in the principle of the positive production of entropy in each local element of a system, whether isolated or not. However, it rapidly became apparent that, as a law, its epistemological status was much less clear than that of the law of the conservation of energy. "It is," Bergson would say, "the most metaphysical of the laws of physics."[7] Classical presentations of thermodynamics associated the law of the conservation of energy (the first principle) and that of increasing entropy (the second principle) but insisted also on their different statuses. The first was a law of nature, an ontology that in effect promoted energy to the rank of substance, in the philosophical sense of that term. The second constituted a law of observation in a descriptive science. It was linked to the specific "boundary conditions" of each problem posed. The second principle, therefore, could not be deduced from the first, whatever Ludwig Boltzmann might have wished.

In 1870 Boltzmann thought he had demonstrated that the elastic collisions between molecules at the heart of a gas invariably produce a heat transfer from the hot to the cold parts and thus an irreversible evolution that he could link to a mechanical quantity H, which, like entropy, varied only in one direction (the H theorem). He soon realized, however, that his demonstration depended on an implicit and unproven hypothesis. This hypothesis, called "molecular chaos," allows the absence of correlations between the velocities of molecules in the initial state of the gaseous mass, even though in the final state such correlations obviously exist between the molecules that have collided. The hypothesis of molecular chaos thus subtly admitted from the start the irreversibility it was supposed to demonstrate.[8]

■ Entropy, Disorder, Information

Boltzmann's theorem provoked perplexity among the physicists of his time. How could a monotonic function, always varying in the same direction over the course of time, be related to a mechanical system when the laws of mechanics that rule its evolution imply no direction of specific evolution for any of the system's dynamic variables? Lord Kelvin was apparently the first to emphasize that the laws of mechanics do not rule out the possibility that at some instant the velocities of all the gas molecules, the evolution of which were supposedly studied by the H theorem, could be reversed. The laws of mechanics imply that the volume of gas must then retrace its path, step by step, through all the states it had previously occupied, as if the film of that sample's original evolution were to be shown in reverse.

All the functions of the system's variables that diminished in the original evolution must then grow, and vice versa. It is therefore impossible that the function H could undergo only a monotonic evolution. This objection is known today as *Loschmidt's paradox*. "All right, go on," a supporter of Boltzmann would have rejoined. "You want to reverse the velocity of all the gas molecules? Do it!" This challenge underlining the *practical* impossibility of preparing systems where entropy spontaneously decreases perfectly exemplifies the spirit of the response first presented by Boltzmann. No, the laws of mechanics do not disallow the possibility that the entropy of a system left to itself will diminish. But such systems do not usually exist in nature: the initial conditions of real systems are such that their entropy grows.

Boltzmann's thinking later evolved and became more precise. Certainly, the law of the growth of entropy, with its temporal irreversibility, cannot be rigorously deduced only from the laws of mechanics applied to the elastic collisions between the molecules of a gas, since those laws are reversible. But the demonstration given by the H theorem is not an absolute demonstration. It is a demonstration in probability. The law of increasing entropy is a statistical truth. Counting the microscopic conditions of a volume of gas leading to a given thermodynamic situation[9] would easily convince anyone that the number of conditions corresponding to thermodynamic equilibrium, of maximum entropy, is very large compared to those corresponding to some other state. Faced with a system whose initial state does not correspond to the thermodynamic equilibrium, the probability that it will approach the state of equilibrium during its evolution is incomparably higher than the reverse. Of course, it is also extremely probable that, in the past and to the extent to which it already existed, this system would have occupied a state close to the state of equilibrium. "Loschmidt's theorem seems to me to be of great importance since it shows how closely the second principle is intimately linked to the calculations of probabilities when the first is totally independent of it," insisted Boltzmann.[10]

In 1890 Henri Poincaré published a theorem of mechanics known as the "theorem of return," which also attacked the H theorem. He stipulated that a mechanical system of finite dimensions that is in a given state at the instant t will infinitely often return in the future to that state or to a state as near as desired to that one. It would be, therefore, certain that the entropy of the system could not always grow, since it must one day or another (and even an infinite number of times in the future) return to its initial value or to one very close to the initial value. The argument was mostly developed by the young physicist Zermelo, a student of Planck, who used it in 1895 in an article leveled against the H theorem, or at least meant to significantly limit its importance.

Boltzmann's response came quickly. "Zermelo's article demonstrates that my works have not been understood," he noted with some bitterness. "However, I must rejoice in this publication as the first proof that my research might be gaining some attention in Germany."[11] Then he once again developed his probabilistic conception of entropy. The decrease of entropy is not excluded, it is only very improbable, given the enormous diversity of the microscopic situations (characterizing, for example, the volume of a gas) that are compatible with the state of thermodynamic equilibrium compared to the rarity of the microscopic situations compatible with a given nonequilibrium state. The average time of return to the initial state (or to an arbitrarily close state) is proportional to the relative rarity of the corresponding microscopic conditions, and thus correspondingly longer as the state considered is farther from the state of thermodynamic equilibrium. For concrete macroscopic systems, such as a mole of perfect gas[12] in conditions far from equilibrium (for example, with an appreciable temperature difference between two communicating compartments of the thermally isolated reservoir in which it is enclosed), this average time of return greatly exceeds the supposed age of the Earth. The theoretical possibility of such a return does not limit the practical conclusion that the system's entropy will increase.

This is the substance of the argument that Boltzmann developed, which introduces the probabilistic interpretation of entropy, whose future fate we know. In many problems necessitating microscopic analysis, the probabilistic definition of entropy[13] is more amenable than its macroscopic definition. But can chance bring elements in line with the laws of nature if it is no more than an apparent randomness due to ignorance? Who could agree to the letter with Feynman's ironic remark in his *Course on Elementary Physics* that "the irreversibility of phenomena effectively observed is due to the very large number of particles concerned. If we could see the individual molecules, we could not decide in which temporal sense the system evolved . . . but if we do not see all the details, then the situation becomes perfectly clear."[14] Shouldn't we rather agree with Poincaré, for whom, if chance could help us predict the future behavior of physical systems, it must be "something more than the name we give to our ignorance"?[15]

The probabilistic interpretation of entropy has, from my point of view, both great merit and one very great difficulty. Its merit lies in its emphasis on the collective, *macroscopic* character of this concept, which makes sense not for an isolated particle (what is the temperature of an isolated particle?) but only for systems having very many degrees of freedom.[16] Its inconvenience lies in the impression it gives that entropy is a concept irremediably endowed with subjectivity, a concept bearing on our relationships to things rather than a quality or a quantity belonging to the things themselves.

It is natural to see entropy as a subjective notion as long as it is expressed as a function of *probabilities*. In Boltzmann's theory, as in the modern presentations of statistical mechanics, the probability of a state characterized by a definite set of values of the system's degrees of freedom is given a priori. The fundamental representation of the theory where microcanonical ensembles are introduced gives an equal weight to each configuration possible in the phase space or each complete set of quantum numbers. Thus, the "probability" of a macroscopic configuration of the system, which is characterized by a given value of the parameters such as volume, pressure, or temperature, is proportional to the number of different ways of realizing such a configuration from the point of view of its microscopic variables (position, orientation, and velocity of each molecule of gas, for example). This rule of a priori equiprobability for all the possible microscopic configurations of the system does not seem to have any justification other than its simplicity and its practical efficacy.

Following Boltzmann's work on the probabilistic interpretation of entropy, the notion underwent a progressive slippage (in fact, several successive changes of paradigms, each of them tending to reinforce the concept's subjective character). Although this slide was continuous during the first half of the twentieth century, three essential stages can be distinguished in which entropy is associated with different notions: following the notion of *probability* came that of *disorder* and finally that of *loss of information*.

An undeniable parallel exists between the growth of a system's entropy and the disappearance of order. If a volume of gas divided among two communicating balloons contains a cold gas on one side and a hot gas on the other, the spontaneous evolution of the system leads to a mixture of uniform temperature. The initial order, in which the hot gas was found on one side and the cold gas on the other, has disappeared. In the same way, a drop of milk added to a cup of coffee diffuses, and this diffusion is accompanied by an augmentation of entropy. The initial "order," the milk on the one hand and the coffee on the other, dissolves into a disordered mixture. The evocation of this parallel between entropy and disorder goes back to Clausius himself. Boltzmann called on the notion of molecular disorder and commented on the idea of the "thermal death of the universe." Little by little in the universe (or at least, said Boltzmann, in the portion of the universe surrounding us and obeying the second law of thermodynamics), hot and cold transform themselves into warm, the rare changes into the common, the heterogeneous changes into the homogeneous, order transforms itself into disorder. The world is thus condemned to a uniformity incompatible with the appearance of ordered structures and individuals such as living beings. This strange apocalypse, which clashes with the popular ideology of scientific and social progress,[17] as well as with the religious idea of finality, frightened the thinkers of the time and inspired the most celebrated science fiction writers of the century's end into despairing commentaries: Camille Flammarion's *End of the World* (La Fin du monde) appeared in 1893.

It seems, however, that the idea of a complete identification between entropy and disorder is fairly recent. Jacques Tonnelat attributes it to Erwin Schrödinger, who in his book *What Is Life?* (1944) proposed taking the inverse of the thermodynamic probability, or the opposite of entropy, as a measure of disorder. But parallelism is not identity, and tempting as it was, the identification of entropy and disorder was not acceptable. As noted in particular by Peter Landsberg,[18] the notion of disorder is an intensive notion, whereas entropy is an extensive notion (the combination of two identical systems must have the same degree of disorder as each of them taken separately, whereas the entropy of the global system is the double of the entropy of each initial system). The recent development of the physics of irreversible processes has amply demonstrated the possibility of imagining situations in which entropy grows even while the evolution of the system is accompanied by an appearance of order: the formation of crystals in a supersaturated solution is just one example.

From the historical point of view, the third stage of the evolution toward a subjective status for entropy is mostly superimposed on the second. It involves identifying the concept of entropy with the notion of loss of information. The idea of a relationship between these concepts goes back to James Clerk Maxwell. In 1872 he imagined a microscopic demon whose intervention, according to him, belied the second principle. Placed at the entrance of the opening between two balloons containing a gas at a uniform temperature, this demon sorts molecules according to their velocity. For example, it lets only the rapid molecules of the reservoir at right pass to the reservoir at left and vice versa for the slow molecules. In this way the demon produces a temperature difference between the two parts of the reservoir initially in a state of thermal equilibrium, thus contravening Carnot's principle. Boltzmann also commented on the relationship between the increase of entropy and the loss of information. In the 1930s the theorist Richard Tolman, the mathematician Emile Borel, and the physicist Leo Szilard were among those who paid the most attention to the relationships between the two notions. But it was only after World War II that Claude Shannon proposed assimilating the two concepts in his work *The Mathematical Theory of Communication*. Finally, in 1948, Leon Brillouin articulated a "generalized Carnot's principle." This principle stipulates not only that the entropy of an isolated system is condemned to growth but also that the difference between the system's entropy and the information acquired on the system is itself condemned to increase. The unit of information he introduced is expressed as $k\ln_2 I$, where k is Boltzmann's constant and I is the information expressed in "bits," corresponding to the minimum number of binary signs by which it can be expressed. With the presence of the constant k, it has the same dimension as entropy.

Thus entropy and information oppose each other and should be subtracted from each other. In fact, it can be shown that, on the one hand, a given quantity of "information" acquired over the course of a physical measurement is offset by the concomitant production of at least the same quantity of entropy (Brillouin says that it is paid for by the consumption of at least an equal amount of negentropy) and that, on the other hand, any information acquired on a system is equivalent to the possibility of taking that system into a more organized state of lesser entropy. In this case, the diminution of entropy is at most equal to this information. For example, to arrange the molecules, Maxwell's demon must first see them, that is to say, it must use a source of light whose characteristic temperature is superior to the ambient temperature; but the light diffused on the molecules it illuminates ends up by being thermalized, and this decrease of temperature entails an increase of entropy. It is possible to show that the information obtained on the velocity of these molecules and the order created by their sorting are at most equal to the negentropy consumed. This double possibility of using negentropy to create information and to then re-create negentropy with the help of this information (up to the limit of the initial expenditure) prevails universally.

The particular ability of humankind to act on matter to organize it is implicated here. We accomplish in daily life what Maxwell's demon accomplishes on the microscopic scale. The effectiveness of our power depends, of course, on the information we possess beforehand on the matter to be organized. The degree of organization or negentropy that we can obtain is always less than the quantity of negentropy we need to spend to obtain this preliminary information.

Brillouin seems thus to have established an equivalency between entropy and information. It is true that by memorizing a written text one can then reproduce it, even if the original is destroyed. Our capacity for organization allows us to create order where disorder reigns. Brillouin's principle, however, sets the limits of such power. Ultimately, information is a means of stocking available negentropy to use later as needed. It allows the entropy's balance of debits and credits to diminish provisionally without reversing it.

Are information, will, memory, and ability to organize of the same nature as entropy? To admit that would be to deny entropy an objective status. We must not forget that the identification of information and negentropy is based on a subjective interpretation of probabilities, on the apparently arbitrary decision to give an equal probability to all the microscopic configurations possible in a system of fixed energy. In physics, the efficacy of such a "recipe" is quite mysterious. It evokes, for Lecomte du Nouy, the feats of legerdemain of the magicians who enchanted him in his childhood. "I learned, later, that the hat had a double bottom and that the rabbit was already in the hat, or in a hidden compartment of the table, but I still haven't understood the 'trick' of statistical laws."[19] The identification of negentropy with information also opens the door to conclusions that are difficult to accept from the strict point of view of physical thermodynamics. Should a different entropy be attributed to two hands in a card game shuffled under the same conditions according to whether one of the hands contains remarkable card sequences or whether the player knows about this result? It is thus no exaggeration to say that Brillouin's principle, the identification of entropy and information, overreaches its objective. The principle could therefore be rewritten in another way, using only homogeneous quantities:

$$\Delta S \geq \Delta S - \Delta \mathcal{S} \geq 0,$$

where $\mathcal{S}$ represents simply the "material" negentropic potential that, following physico-chemical modifications in the central nervous system, appears somewhere in the brain when it becomes aware of a measurement or information, and not the subjective information itself. It is understood that $\mathcal{S}$ indeed takes a value close to $k\ln_2 I$, as Brillouin wanted, but within limiting conditions that remain to be analyzed.[20]

■ Does the Growth of Entropy in Familiar Phenomena Have a Cosmic Origin?

Boltzmann's pioneering work has shown the law of increasing entropy to be statistical, valid exclusively for large systems, or even more, for ensembles of systems. The next step consists of establishing both a statistical and objective interpretation of entropy. One apparent advance in this direction was accomplished recently by the American physicist David Layzer, who suggested linking the statistical properties of entropy to those of cosmological chaos. This still nebulous idea may become more precise in the near future as we better understand the global mechanisms of cosmogenesis and cosmological expansion. It is not new, however. Its origin can be traced to Emile Borel for its mathematical ingredients and to Arthur Eddington for its cosmological flavor.

The idea begins with a critical analysis of the function of Maxwell's demon. Sorting the gas molecules, the demon can take advantage of local heterogeneities in the velocities of the molecules, heterogeneities that are un-observable at our scale. Apart from that advantage, it acts at its microscopic scale exactly as we would at ours, for example, in sorting tennis balls. In the example of diffusion between two balloons filled with gas at different temperatures, the growth of entropy thus appears less a problem of disappearing information than one of changing the *scale* of information. This applies also to the example of a drop of milk dispersing in coffee. Globally, at our scale, information does indeed disappear. We can no longer distinguish the two constituents in the café au lait. But a minuscule organism living in the coffee would observe the separation of the drop of milk into minuscule globules and would continue to distinguish these globules from the surrounding coffee. The possibility would always exist of separating the two constituents. The total volume of the globules would still be exactly equal to the volume of the initial drop of milk. The information has not changed. It is only when an approximate description is accepted, one founded on a subjective notion of global appearance, that the information necessary for the description of the observation changes. The information thus depends on the scale pertinent to the description. Of course, this change in scale is itself difficult to analyze in objective terms. Where should one place the limit between the macroscopic world, in which the information is accessible or even usable for the purposes of organization, and the microscopic world, where the information would be inaccessible? Does this limit depend on the power of the available instruments of investigation?

Put in Borel's terms, the growth of entropy thus translates the universe's tendency to move toward a state with a progressively finer structure. "The evolution of the Universe might thus be conceived of as tending to produce a more and more complicated state, unable to be perceived and used by other than smaller and smaller beings."[21]

The only truly objective process of increasing entropy, if it existed, would therefore be associated with the total disappearance of information and not with its transformation to a smaller scale. In the same way, the only process of increasing negentropy that would be truly objective would have to be linked to the absolute appearance of a quantity of information and not to its change in scale.

The universe itself might well constitute this source of objective information. According to Layzer, a mechanism for the objective production of negentropy is seen in cosmological expansion as general relativity explains it. This claim, however, is argued only in the framework of a particular model, one that obeys the "strong cosmological principle." This principle states that the distribution of matter and energy at the heart of the universe results from pure chance and shows evidence of neither particular place nor direction. Local irregularities, such as the galaxies and the stars of which they are constituted, are no more than fluctuations at the scale of the universe, reflecting the essential fluctuations of quantum origin that presided over the appearance of matter in the very first instants of the Big Bang. These fluctuations escape all determinism by their very nature, and the matter that today forms the universe can thus obey this statistical principle. Finally, the model allows that the universe is without a frontier. In fact, according to general relativity, the universe might, under certain conditions, be finite but unlimited, in the true sense of this term. If this were the case, there would be neither a separation between a part filled with matter and an empty part, nor a "center," nor any other privileged place. To return to the established image, it would be like the surface of a balloon and could thus be traveled endlessly without ever encountering a limit.

This model strips the universe as a whole of any objective order, thus precluding any objective description, even a partial one. For example, attempting to describe objectively the solar system's situation in the universe would be deluded, since there certainly exist a great number of planetary systems of the same description, however detailed, in the myriad of galaxies.

Briefly, Layzer's hypothesis denies the possibility of information or objective order in the universe at this scale of description. All appearance of order or negentropy at a lesser scale is thus a *creation*. Only the creation of negentropy at a lesser scale, if it were to exist, could be called objective.

The mechanism proposed for the local creation of negentropy involves the two stages of the expansion of the universe: the cooling of matter and light and the gravitational collapse of clouds of cold matter. The cooling that accompanies the expansion of space explains why the characteristic temperature of background radiation, the fossil vestige of a concentrated and hot universe, is today no more than 2.7°K. But this cooling has a different effect on massless photons and material particles. The temperature of the first decreases proportionally to R^{-1}, where R is a characteristic parameter that might be called the radius of the universe, whereas the temperature of the second decreases as R^{-2}. A permanent source of negentropy is thus created, the corresponding flux being born from the processes coupling matter and radiation. The gravitational collapse of cold material particles, provoking the reheating of gas clouds and their condensation into stars, plays a central role among these processes. In allowing the formation of the Sun, they gave to the biosphere and to humanity the negentropy necessary for them to survive. "The general struggle for existence of animate beings is therefore not a struggle for raw materials—these, for organisms, are air, water and soil, all abundantly available—nor for energy, which exists in plenty in any body in the form of heat (albeit unfortunately not transformable)," Boltzmann was already writing more than a hundred years ago, "but a struggle for entropy, which becomes available through the transition of energy from the hot sun to the cold earth."[22]

In terms of the hypotheses presented, the link between the growth of entropy in phenomena on the smaller scale and the founding cosmic process remains to be better and more precisely understood.

First let us remember that the phenomenon of increasing entropy is specific to the collective. One atom, even radioactive, has no age: it always has the same probability of existing the following minute. On the other hand, a collection of radioactive atoms, a sample of radioactive matter, allows the construction of a thermodynamic clock. The more massive the sample, the more precise the clock. The thermodynamic duration seems, therefore, to be infused in matter with a degree of precision that depends on the size of the studied sample. David Bohm proposed the term *implicate order* to designate a situation in which an external order is infused into matter.[23] He suggested an analogy for it. The temporal order would be infused in matter as the geometrical order of an object's surface is infused in the photographic plate of a hologram.

A hologram uses the interference of two laser beams, one of which has been diffracted on the surface of an object, to produce a very complex figure that depends at every point on the object's overall form. This external order does not appear on examination of the photographic plate in ordinary light. It is, however, immediately perceptible when viewing the hologram with a laser of the same frequency as the one used to construct it. The geometrical form and even the relief of the object photographed is then seen to appear. It must be noted that the detail and resolution of the reliefs depend on how much of the hologram the laser illuminates. If the region illuminated is limited, the object still appears in its entirety, but with less precision.

In the universe, the external structure to be "implicated" is cosmological duration, the arrow that universal expansion gives to time. The "holographic" support in which this order is "implicated" is the matter it contains. In this analogy, the arrow of time would be reflected, infused, or translated in matter at the level of macroscopic phenomena by way of the law of increasing entropy with a precision that increases with the size of the system being considered.

Such an analysis suggests a link between the "small systems" and the universe. This link had already been invoked by Eddington, who in 1935 expressed the idea that the growth of entropy in phenomena at our scale could be linked to cosmological expansion.[24] The mechanism for this is not yet clear, but the probable conclusion is that the growth of entropy in concrete systems usually takes place in two stages. First comes a "subjective" stage, in which it is linked to the change in scale of the information available on the system, which passes progressively from the macroscopic scale (a drop of milk for example) to the microscopic scale (globules). At this stage, what an untrained observer would see as an increase of entropy could be considered as an isentropic[25] process by an observer using the most precise instruments. In a second stage, the microscopic information dissolves and disappears through the weak perturbations that the rest of the universe induces on the system. The conclusion that entropy is increased thus becomes universal and objectively founded.

What are these weak perturbations that join every system to the rest of the universe? Does gravitation play a decisive role here? That was the opinion of Borel, who pointed out that moving by a centimeter a gram of matter situated as far away as Sirius would cause a change in gravitational potential that would profoundly alter, in less than a microsecond, the microscopic state of a volume of gas in a terrestrial laboratory. In the current state of the theory, it is too early to reach a decision on this subject.

■ ■ ■

Thus the difficulty of the debate on the objectivity of increasing entropy could reside in the simple fact that until now we have neglected to distinguish two successive stages in this process, one anthropocentric and subjective, the other cosmological and objective.[26] The final objectivity of the growth of entropy would depend on the ultimate disappearance of information due to the truly chaotic properties of the cosmos. We are not yet, however, entirely sure of this, and we still know little about the coupling mechanisms responsible for duration. But we are perhaps on the eve of great revelations, on the eve of worming this secret of time out of nature. The key that Boltzmann so long and vainly sought, to such a point that his despair contributed to his suicidal depression, was perhaps well hidden beyond the clouds . . . in the red shift of galaxies, which was not discovered until twenty years later, and in the original chaos, which we still do not fully comprehend.

14

Dissipative Structures, Cyclic Reactions

During the twentieth century, thermodynamics was developed to study the dynamic behavior of nonequilibrium systems. Two groups of systems were distinguished, those systems near equilibrium, when the factors keeping them from reaching that state are not too great, and those that, on the contrary, are kept far enough from equilibrium that nonlinearities, notably feedback loops, appear in them. This latter case is particularly interesting, in that the theory (developed mostly by Prigogine and the Brussels school) shows how, in the systems' attempts to return to equilibrium or to stay as close as possible to it, they can acquire and maintain remarkable structures. To do this, it is essential for their entropy to flow toward the exterior, a flow that must be greater the farther they are from equilibrium. Simple examples of such systems, called dissipative systems, have been studied as much theoretically as experimentally, especially in chemistry. Spatial regularities (ordered figures) or temporal recurrences (cycles) often appear in them. These systems furnish a model for understanding how, while obeying the laws of thermodynamics, especially the law of entropy growth, highly nonlinear systems such as living organisms are able to construct and maintain themselves.

■ Development of the Thermodynamics of Irreversible Processes

Clausius, Boltzmann, and Gibbs developed a theory of thermodynamics describing, above all, the state of equilibrium toward which all physical systems tend. To extend the theory to the description of systems in disequilibrium, the following generation of thermodynamicists (Constantine Caratheodory, Pierre Duhem, Theophile de Donder, Lors Onsager, Ilya Prigogine) proposed first generalizing the Clausius-Carnot law expressing the ineluctibility of increasing entropy in isolated systems. To do this, they distinguished two contributions to the entropic menu of nonisolated systems: entropy *created* at the heart of a system, and the flow of entropy through the system's outer layer (from the system toward the exterior or vice versa). According to Clausius's intuition, irreversible processes are characterized by a local positive source of entropy, whereas reversible processes are characterized by a local null source. The generalized law of Clausius-Carnot is thus written in the form $dS = d_iS + d_eS$, where $d_iS \geq 0$ represents the local source and d_eS represents the outward flow.

For isolated systems this new principle is indeed equivalent to the old, since the term of flow toward or from the exterior is then null, thus reducing the preceding equation to $dS \geq 0$.

In addition, the thermodynamics of irreversible processes are based on the hypothesis of "local equilibrium," according to which all systems in disequilibrium can be divided into elementary subsystems that allow each of them to be considered as momentarily being in equilibrium. The laws of classical thermodynamics can then be applied at the level of these subsystems. This hypothesis makes it possible to calculate the value of the thermodynamic functions (and entropy in particular) of the whole system, even if this system is manifestly not in equilibrium. This hypothesis of local equilibrium is, however, not valid in describing the most violent phenomena, such as those involving the formation and the propagation of shock waves and so on. It is, however, very useful in describing current situations of nonequilibrium.

The theory initially was developed for and applied to stationary systems close to a state of equilibrium. In this state, entropy takes on the largest possible value and remains constant over time, so that the rate of increase in entropy is null. It has been demonstrated that systems in which the imposed constraints impede the establishment of equilibrium, but which may nevertheless come close to that state, produce entropy at a rate that takes on the minimum value compatible with these constraints. This is the *theorem of minimum entropy production* (Ilya Prigogine, K. G. Denbigh). The field of validity for the thermodynamics of irreversible processes thus defined is that of "linear thermodynamics."

As long as the system remains in a state of nonequilibrium, however, the production of entropy is never null. A stationary state out of equilibrium thus has an internal physico-chemical activity that produces entropy. For "entropy," which is a state function of the system, to remain constant, which the very idea of a stationary state demands, the generalized Clausius-Carnot law (with $dS/dt = 0$, but $d_iS/dt > 0$) requires that $d_eS/dt < 0$: entropy must flow toward the system's exterior. In other words, isolated systems cannot attain such stationary states. On the contrary, to do so the systems must be open to the exterior, disgorging entropy there or, in other words, consuming "negentropy" present in their environments. This remark sheds light on, in particular, the most banal observations of biology—every complex biological system, being out of equilibrium, is condemned to degradation and death once it is isolated.

Stationary states close to equilibrium are stable states, because every deviation of a system-state variable leads the system into a state for which the production of entropy is larger and because the system tends spontaneously to adopt the state associated with a minimum production of entropy. Thus any disturbance momentarily applied to a stationary state does not modify the state except in a transitory way. As soon as the perturbation has ceased, the system resumes its previous state. In fact, for all phenomena proper to linear thermodynamics, the system remains in the state of equilibrium's zone of attraction and tends to get as close to it as possible. If the constraints imposed on the system and keeping it from attaining the state of equilibrium are dropped, the system spontaneously and invariably evolves toward equilibrium.

But the thermodynamics of states of equilibrium and, in a large measure, the linear thermodynamics of irreversible processes do not account for certain elementary observations. They have, in the end, always left the physicists and the philosophers of science unsatisfied. Basically, they do not explain how, in a world dominated by the notion of the state of equilibrium, itself characterized by entropy and a maximum disorder, the "scandal of life" came to be produced.

The problem of life is found, of course, in the background of the thermodynamicists' concerns.[1] There remain, however, elementary observations belonging without dispute to the field of the inanimate world that do not accord well with thermodynamics as I have been developing it here. Spontaneous manifestations of spatial or temporal order collide with the idea of a world drawn towards homogeneity and disorder and indicate the limits of linear thermodynamics.

■ Some Examples of Nonlinear Chemical Reactions

Putting a drop of mercury onto the flat bottom of a bottle of hydrogen peroxide with a pH previously set at 7 produces a release of oxygen with a marked rhythm, while the drop seems alternately to expand and contract. This last phenomenon gave the experiment the name of "Bredig's heart" (Bredig and Weinmayr 1903). Another periodic reaction was noted by W. C. Bray in 1917 concerning the decomposition of hydrogen peroxide in the presence of iodate of potassium and sulfuric acid. This reaction is marked by a regular variation in the concentration of iodine and a rhythmic release of oxygen.

Other examples lead not to a rhythmic activity but to a spatial structuration. Placing a drop of a solution of silver nitrate on a layer of gelatin impregnated with potassium bichromate produces superb concentric rings of silver precipitate, which form a figure that recalls to some extent Newton's rings in optics but that is unexpected in a chemical reaction between homogeneous bodies governed by the laws of linear thermodynamics. This observation was noted, apparently for the first time, by R. E. Liesegang as early as 1896. Another example among many is furnished by Zabotinsky's reaction (1964). It consists of the oxidation of malonic acid by potassium bromate in the presence of the ions Ce^{3+} and Ce^{4+}. This reaction is followed by a spatial structuration in the test tube in which it is conducted: layers alternately dark and light appear throughout the entire tube height.

Such systems do not necessarily contradict the law of increasing entropy. First, as already seen, there is no absolute equivalence between entropy and disorder.[2] Consequently, the appearance of a spatial order can be accompanied by an augmentation of the system's entropy, as is the case with crystals appearing in a supersaturated solution. Moreover, without referring to this distinction, it is not necessary that the systems evoked produce negentropy, since these systems are not isolated: they can thus profit from a flow of negentropy from the outside. This flow can even be superior to that which would be strictly necessary to maintain them in a stationary and homogeneous state. For example, a system oscillating in time must have a total entropy alternately growing and lessening, in unison with the appearances of the phenomenon. Consequently, these systems certainly do not obey the law of minimum entropy production, according to which the outward flow of entropy must be constant. They do not, therefore, belong to linear thermodynamics' field of application.

Prigogine verified these conclusions by developing, on the theoretical and experimental levels, the nonlinear thermodynamics of irreversible processes. When the "forces" maintaining a system in disequilibrium are not proportional to the associated "flows," which tend, in linear thermodynamics, to minimize the production of entropy, the state of equilibrium ceases to be the only stable or metastable attractor that the system can reach. Many of these attractor states show a high degree of spatial organization or a marked course of evident rhythms. Such situations are frequent when several chemical reactions take place simultaneously within the same system, each borrowing its reagents from the others.

■ **Coupled Chemical Reactions Can Lead to Cycles**

The Brussels school formulated the theory of the behavior of an open and complex system that Prigogine called the "Brusselator,"[3] which consists of a particular series of four coupled chemical reactions. It is possible to show that the system of equations ruling the kinetic behavior of such a system predicts, in certain conditions, a periodic variation of the concentration of certain intermediary products.

In France in 1975 Adolphe Pacault and his team constructed a true chemical clock, whose hands are moved by a chemical reaction with a particularly well stabilized period.[4] Such a clock illustrates anew the idea that measurements of time can rest on two distinct categories of physical phenomena—those that arise from mechanics or from the principle of least action, on the one hand, and those that arise from thermodynamics, on the other.

Numerous other examples of oscillating reactions obtained in coupled chemical reactions have been studied since, both theoretically and experimentally.[5]

■ ■ ■

To summarize the situation, the stationary or steady conditions attained by nonequilibrium thermodynamic systems are generally states of a local minimum of entropy production. The transitory changes of exterior constraints and internal fluctuations affecting the system, if they do not go beyond a certain critical value, are purely and simply reabsorbed without changing the system's overall state. If these changes surpass a certain critical value (a threshold value that depends on the state considered), the system can leave the stationary state and not return to it. Several attractor stationary states can then exist with the same exterior constraints. The particular state toward which it effectively tends can then depend on experimentally uncontrollable internal fluctuations.

The nonlinear thermodynamics of irreversible processes helped explain the conditions for the appearance of dissipative structures. In the case of chemical reactions, this phenomenon involves systems of coupled chemical reactions appearing in open systems and possessing autocatalytic or cross-catalytic effects.[6] These catalytic effects are responsible for nonlinearities between "forces" and "flows," between "affinities" and "reaction rates." The metabolic mechanisms of living beings are noted to be very much of this type. Prigogine writes on this subject:

> This catalytic character of the chemical reactions needed to lead to an instability is particularly interesting when one thinks of the principal biochemical reactions (such as the cycle of glycolyse). All these cycles are maintained thanks to a very complex network of regulations; all the decisive stages are ruled by mechanisms of activation or of inhibition, or even of the two simultaneously.

The most general conclusion that can be drawn from these studies is that, while disorganization and inertia are the norm in situations near to equilibrium, beyond the threshold of instability the norm is auto-organization, the spontaneous appearance of an activity differentiated in time and space. The forms of this dissipative organization are very diverse. Certain systems acquire a spatial inhomogeneity in a spontaneous manner, others adopt a periodic temporal rhythm, true chemical clocks, others even associate spatial and temporal structurations. Finally, some acquire veritable natural borders, of dimensions determined by the parameters characterizing the activity of the system.[7]

The field of dissipative structures opened the way to a comprehension of phenomena in which structures spontaneously appear. It constitutes a new proof of the need to reintroduce a historical dimension into the description of material phenomena, since the possible multiplicity of stationary or oscillating states accessible to a given system, in the given conditions of constraints, and the fact even that these states each possess a different zone of stability introduce the characteristic phenomenon of hysteresis. The system's behavior for a given value of external constraints depends not only on the external conditions but also on its own history. Such a behavior could not be observed in the linear field of thermodynamics.

We must now leave inanimate phenomena to seek to expand the new knowledge acquired into the area of life. Despite formidable differences in complexity, certain examples of cellular organization irresistibly evoke the conditions for the appearance of dissipative structures.

From the Time of Things to the Time of Living Beings: A Leap in Complexity

The examples of dissipative systems borrowed from the field of chemistry involve only a few chain reactions, such that the reaction products of certain stages influence the progress of other stages. The temporal cycles engendered by these simple models suggest an analogy with the living world, in which temporal cycles of all kinds play a large role. Can a rhythmic behavior of living matter—for example, in glycolysis, respiration, or even photosynthesis—be explained by applying the laws of the thermodynamics of irreversible processes to the open systems that are living organisms? Prigogine thinks so[1] and wrote about it as early as 1946. According to him, far from prohibiting them, the laws of thermodynamics can enforce highly organized behaviors on certain chemical systems.

We must not underestimate the formidable leap in complexity between simple, known chemical models involving at most ten or so different compounds and the functioning of the smallest bacterium, which simultaneously involves billions of components in a highly organized structure. Could this vast distance between the simplicity of an artificial system and the complexity of a living system involve a difference not only in scale but also in nature, namely, the emergence of "auto-organization," a property specific to the living world?

■ The Chemical Complexity of Living Systems: The Example of Glycolysis

Before being able to classify the phenomena of life among dissipative structures, it must be verified that they obey the three conditions imposed by the laws of thermodynamics. First, living systems must be able to decompose into a large number of simple units susceptible to short-range forces of interaction, like Boltzmann's gas molecules. Second, these systems must be of the "open" type, that is, able to exchange energy and matter with their environment. Finally, these systems must be maintained, with a strong external constraint, in a state of nonequilibrium and far enough from equilibrium to escape the domain of linear thermodynamics. For this to be true, their evolution must be directed by a chain of interactions involving feedback loops or other nonlinearities, as with the chemical reactions already cited.

These conditions are generally met by living systems. Let us consider a bacterium, a living being constituted of a single cell. Its cytoplasm encompasses millions of molecules of different proteins of varied form and composition that are able to interact in their mutual collisions. The cell is an open system due to the permeability of its membrane. Also, some of its organelles (e.g., mitochondria and chloroplasts) specialize in exchanges of energy with the outside, such as respiration or the capture of solar energy. These exchanges of energy and matter maintain the cell in a state characterized by a very large structural heterogeneity, far from thermodynamic equilibrium. Finally, the ordinary exercise of the vital functions in this cell brings into play chains of complex reactions, where retroactive stages, with autocatalysis or cross-catalysis, are frequent. The enzymes, which are proteins whose synthesis DNA directs at the heart of the cell, play the role of catalysts in these reactions. Their active function is generally linked to the extremely convoluted form of their molecules, which form niches or traps designed to fit with parts of other molecules with which they are destined to interact, thus facilitating their meeting. The cellular structure also contains numerous semipermeable membranes that define not only the contours and the internal machinery of the different organelles but also the reticular network, the internal skeleton. These membranes also maintain certain constraints inside the cell. They impede inopportune mixtures and contribute to maintaining the cell in a nonequilibrium state.

Having admitted that living organisms fulfill the conditions of application for the thermodynamics of irreversible processes, two questions can be asked. Can this theory account for the organization and maintenance of vital spatial and temporal structures necessary to the particular functions of living systems? Can it furnish us with information on the appearance of life, on the passage from the inanimate to the animate?

Let me illustrate the first question with an example chosen for its importance in the biosphere: the cycle of glycolysis, that is, all the reactions by which animal cells use their reserves of glycogen to fabricate adenosine triphosphate (their immediate energy source, or "fuel").[2] Glycolysis is no more than an archetype among the vital functions. Its mechanisms are relatively well known,[3] unlike other vital functions such as chlorophyllous photosynthesis, which has some stages that still are not completely elucidated (fig. 11).

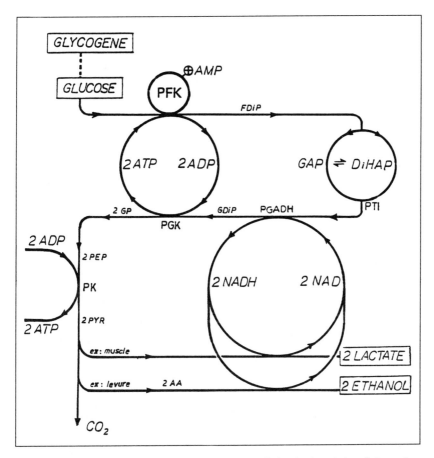

Figure 11. Schematic drawing of the reactions taking place during the degradation of glucose into lactate (in the muscles) or into ethylic alcohol (in yeasts). This process allows the transformation of two molecules of ADP into molecules of ATP, which is more energetic. The highest loop, catalyzed by the enzyme phosphofructokinase (PFK), which is itself sensitized by the presence of the precursors of ADP, is highly nonlinear and appears to be the principal agent responsible for establishing periodic oscillations in the overall course of the reactions. From A. T. Winfree, *The Geometry of Biological Time* (New York: Springer-Verlag, 1980), p. 285. Used by permission of Arthur T. Winfree.

The degradation of glucose to fabricate molecules of adenosine triphosphate (ATP)—two molecules of ATP for each molecule of glucose from two molecules of adenosine diphosphate, or ADP—is one of the routes by which living organisms meet their energy needs. The principal stages of this synthesis of ATP have been known since the beginning of the century, but it was not until 1957 that A. J. Duysens and J. Amesz observed the oscillations that usually accompany this synthesis. J. Higgins in the United States and E. Selkov in the USSR were the first (1967–68) to propose mechanisms to explain these oscillations.

The principal stages of this synthesis of ATP are shown in figure 11. This figure looks at only one part of the energetic cellular workings, since at this point the synthesis, storage, and degradation of the glycogen into glucose have already taken place, whereas after this stage, the resulting lactic acid and ethanol can undergo even further degradation in an aerated environment.

Glucose is first phosphorylated (through the action of ATP) and then transformed into phosphorylated fructose. This phosphorylated fructose receives a second atom of phosphorus, once again from an ATP molecule, during a reaction catalyzed by an enzyme that is particularly important here, phosphofructokinase (PFK). The product of the reaction, fructose diphosphate (FDiP), is then split into two trioses (each with three carbons rather than the six of glucose and fructose), D-glyceraldehyde phosphate (GAP) and dihydroxyacetone phosphate (DiHAP). These two compounds transform each other under the influence of another enzyme, phosphotriose isomerase (PTI). The DiHAP is then oxidized by NAD (nicotinamide adenine dinucleotide) in the presence of phosphoric acid and under the catalytic action of a specific enzyme, phosphoglyceraldehyde dehydrogenase (PGADH), in such a way that the trioses are finally transformed into glyceric diphosphate acid (GDiP), while the NAD is reduced to reduced nicotinamide adenine dinucleotide (NADH). The molecules of GDiP are then themselves transformed, under the action of ADP coming from the reaction catalyzed by the phosphofructokinase and in the presence of the enzyme phosphoglyceratekinase (PGK), into glyceric phosphate acid (GP), while the original ATP molecules are regenerated. Thus, during the course of this cycle (represented by the circle at the top and to the left of the figure), the second atom of phosphorus, attached to each triose, is returned to the ATP and the trioses are transformed into monophosphates. After a further transformation, the phosphoglyceric acid is transformed into phosphoenolpyruvate (PEP). During the reaction that follows, and catalyzed by the pyruvatekinase enzyme (PK), the phosphenolpyruvate gives its remaining phosphorus to a molecule of ADP, transforming it into ATP while itself giving birth to a pyruvate (PYR). In some organisms—mammalian muscular tissue, for example—the molecules of pyruvates are themselves reduced, under the action of the NADH previously formed, into lactate. This product provisionally accumulates in the cell and causes, in overexerted muscles, the cramps and muscular pains with which we are all familiar. In other organisms, such as the alcoholic yeasts, the pyruvate is split into acetaldehyde (with a release of carbonic gas). Finally, the acetaldehyde is oxidized by the NADH into ethanol, which is nothing more than the alcohol of fermented drinks.

In this ensemble of reactions catalyzed by phosphofructokinase, the important point is the extreme sensitivity of this enzyme to the presence of adenosine monophosphate (AMP), itself a precursor of ADP.[4] AMP can fix itself to certain particular sites of the enzyme, which causes a change in its configuration multiplying by one hundred its activity potential. Thus, the production of ADP during this reaction tends to accelerate the action of the phosphofructokinase by an autocatalytic process of multiplication. The reaction accelerates until, in the following stages, the ADP product is itself degraded, causing a deceleration of the reaction. This mechanism thus produces the oscillations whose existence can be observed, especially through the variations in fluorescence accompanying the variations of the concentration of NADH in the organic environment. Under most conditions of observation, the period of the oscillations is about one minute.

The cycle of glycolysis and the generating mechanism of the oscillations illustrate how much more complex the functions of living organisms are compared to the simple mechanisms of the cyclic reactions studied in the laboratories of mineral chemistry. Nature was, in a certain way, "obliged" to invent these sophisticated recipes to take better advantage of the environment. The energetic output of glycolysis, that is, the relationship between the energy theoretically available (by the total oxidization of glucose) and the energy truly at the disposal of living beings in the form of ATP, is—in the favorable case of living aerobic beings—considered to be on the order of 30 percent, a number rarely surpassed by man-made machines. In another vital function, the chlorophyllous synthesis, the mechanisms' complexity is required because the energy of solar photons is insufficient to provoke the synthesis of carbohydrates directly. The successive chemical reactions allow the system to accumulate the energy of several photons, a total of eight for each atom of carbon extracted from the atmospheric carbon dioxide and fixed in a molecule of cellulose. The overall yield of the corresponding reaction is no more than a few percent, but it achieves through this play of helpful intermediate stages what could have a priori appeared impossible.

Now, on to the second question. The thermodynamics of irreversible processes renders intelligible the existence of a structured system whose organization maintains itself despite possible fluctuations of the constraints imposed by the environment. This organization frequently involves a periodicity in the behavior of such systems. On the other hand, the thermodynamics of irreversible processes does not explain how the gigantic tree of life, whose every twig seems more complex than the branch that gave it birth, came to be developed over the course of terrestrial history. It explains the existence of complex systems and explains why these are spatially and temporally organized, but it does not seem able to explain immediately why living organisms climbed the ladder of complexity with such persistence, above and beyond the accidents of a three-billion-year-old history.

■ Complexity and Energetic Integration of Living Organisms

The formidable complexity of living systems is underlined in ways other than just the number of their constituents or chemical stages. In most physico-chemical systems, the exchanges of energy between the system and the outside are relatively homogeneous. They are all of the same chemical, electrical, photic, etc., nature. In living systems these exchanges of energy are varied and interdependent. Thus, the activity of chlorophyllous photosynthesis rests on the contribution of luminous energy (photons), but this is used first to effect transmembranic transportation of electrons, which ultimately allow the production of chemical substances with a high energetic value. The higher up the ladder of complexity of the living world, the more interdependent networks exist in which the energy is transported into a state of less and less entropy, more and more "organized." At this stage it is better to speak of the transport of "information." The chemical structures are thus taken to the level of a code (the genetic code, for example), of an intracellular message (RNA messengers), and then to the level of an intercellular message (hormones or hormonal network). In addition, certain living cells use excitable membranes to transmit electrical messages. In the neuron, propagation of a wave of electric potential along the membrane constitutes the so-called nerve impulse. The transmission of the nerve impulse from one cell to another is usually mediated by a chemical message, due to the liberation of a neurotransmitter in the area of contact between the two neurons, which is called the synaptic space. The nerve impulse can also run throughout the entire nervous system and be directed according to an internal plan toward specific targets. The systems of higher organisms are so integrated that there clearly exist feedforward and feedback interactions between the nature of the nerve messages transported and the physiological structure of the organs, a bit as if the type of program submitted to a computer could influence the machine's structure, as if the software could influence the hardware!

■ From Autopoiesis to Autocomplexification

Energetic integration is an important aspect of the processes that end by giving identity and durability to living organisms. Humberto Maturana and Francisco Varela propose underlining their significance by calling these processes "autopoiesis," that is, the production of living beings by themselves.[5] But this property seems to characterize complex systems in general rather than living beings in particular. We now know how to confer on computerized automata (or robots) the ability to structure themselves and produce copies of themselves. The new property of living organisms, absent or inoperative in artificial physico-chemical systems, would instead be that of autocomplexification. Living organisms draw on themselves and their interactions with the outside for the resources necessary not only to reproduce beings that resemble them from generation to generation but also to bring to light beings that are more and more complex. They achieve this in two stages. First, they structure themselves in a hierarchical manner by spontaneously creating semipermeable barriers (such as cellular membranes) so as to define relatively autonomous levels of organization and circulation of energy and entropy. Many of these permanently intercommunicating subunits merely replicate an archetype and furnish the organism with a reserve of "redundancy." Second, once the degree of redundancy acquired is sufficient to ensure their survival, living organisms use outside strokes of luck and accidents to transform a part of this internal redundancy into new organization and thus climb the ladder of complexity.

We are far from the enthusiastic and naïve speculations of the first cyberneticists, who hoped, armed with the single principle of retroaction, to make their "robots" models of the living world, even of the thinking world. Some of them thought that by applying the paradigm of retroaction coupled not to chemical reactions but to electromagnetic servomechanisms they might understand and even imitate the functions of the brain. This research is, of course, at the origin of the invention of computers. Its logical extension has brought astonishing results, for it has allowed the development of "artificial intelligence"; it has furnished us with machines specialized in "pattern recognition" and those that have the logical infallibility and encyclopedic memory of "expert systems." It is dizzying to see these performances and to think back on the predictions of von Neumann, who prophesied as early as 1948 that "soon, the constructor of robots will be as disarmed before his creation as we are before complex natural phenomena!"[6] Over the past few years certain types of robots have passed another threshold by acquiring two new properties that they will henceforth share with living organisms. They have acquired resistance to minor breakdowns of components and have gained the ability to learn by progressively adapting to a determined function. But it would be excessive to compare these machines, whatever their performances, to living organisms. In truth, the robots in question represent an intermediary degree of complexity between the dissipative chemical systems and the living biochemical systems. For this reason, they furnish priceless models for study.

Part 4

Time, the Engine of Life

What Is Life?

Science is one of philosophy's tools, but it advances only one small step after another. The small "nos" it reveals to us are nevertheless precious. If the large questions remain essentially the same, at least the terms in which they are asked are becoming more and more precise. For example, we still do not know how to define consciousness. The first robot that turns to its creator to declare that it is conscious will cause a great deal of confusion, for self-consciousness has no definition that can be applied to and verified in others. But do we at least know how to define life? The answers that science brings us allow us to better delimit life's characteristic properties. The small "nos" that experimentation provides lead us, progressively, toward a new interpretation of biology in which time might indeed play a central role, just as it did in physics following Galileo's discovery.

■ The Criteria of the Living

A first definition of life is one that opposes it to its ineluctable opposite, death. But is this opposition truly well founded? Is it true that all living beings must die?

It is not absolutely certain that death is the inevitable outcome of life. Some simple beings, such as sea anemones, do not seem to grow old and enjoy a longevity apparently limited only by accident. The record for observed agelessness belongs to the colony of sea anemones harvested in 1862 by Anne Nelson for the aquarium of the University of Edinburgh and preserved under constant surveillance, without change or apparent aging, for more than eighty years. The sea anemones were abandoned and accidentally perished during World War II.[1]

Each living organism is extraordinarily complex, and this complexity appears to be closely related to life itself. An organ like the human brain contains as many cells (about 500 billion) as the number of stars in a galaxy like ours! And these cells are in a close relationship with one another. Could the definition of life have something to do with this complexity? No, because complexity is not a synonym for life. The most complex machines are not necessarily living machines. We could build hypercomplex machines capable of reproducing themselves without them being living machines.

In his book *Chance and Necessity,* Jacques Monod specifies the terms in which the question of life's distinctive character must be asked. First he admits—in agreement with Prigogine—that the auto-organization of living systems, far from defying the laws of thermodynamics, uses these laws: "spontaneous structuration (in living beings) must rather be considered as a mechanism," he writes.[2]

Beyond identifying this preliminary property, he distinguishes three characteristics in life: *teleonomy,* that is, the appearance of having a goal (ensuring the perpetuation of the species) linked to the existence of a genetic program; *invariant reproduction,* that is, the reproduction by an individual of another individual essentially the same as itself—or better, on the molecular level, the identical reproduction of DNA in cellular reproduction; and finally, what might be called *"closure,"* that is, the production of every living being through the sole action of the biological forces animating a progenitor of its own nature.

These three characteristics are of unequal importance. Invariable reproduction, which constitutes one of life's most spectacular characteristics, has often been cited as *the* distinctive characteristic, par excellence, of life. In fact, this characteristic is not specific to the living world. Monod himself observed that in one sense the crystallization of sea salt is a reproduction of like by like. The crystals grow, through successive captures of sodium and chlorine ions from the ambient surroundings, in a geometric pattern furnished by the seed crystal. This definition would require us to characterize as living certain viruses that are no more than simple macromolecules in geometrical forms, which our chemists know how to obtain in a crystallized state. This observation must be qualified, however, because an isolated virus reproduces only by parasitizing living cells. In this sense, it could be said that they possess life only in a latent or potential state.[3]

■ Does Life Have a Plan?

Since reproduction of like offspring alone does not constitute a distinctive criterion of the living, the two other proposed criteria must now be examined: teleonomy and closure. What is the "goal" Monod evokes? In fact, we must address the appearance of a goal since, hypothetically, there is no final cause to direct the course. H. Atlan summarizes Monod's thoughts in this way:

> A teleonomic process does not function by virtue of final causes even if it seems to, even if it seems oriented toward the realization of forms that will appear only at the end of the process. That which determines it (the teleo-

nomic process), in fact, is not those forms as final cause, but the execution of a program, like a programmed machine whose workings seem aimed at the realization of a future state, when it is in fact causally determined by the sequence of states the preestablished program made it pass through. The biological program itself, contained in the characteristic genome of the species, is the result of long biological evolution where, under the simultaneous effect of mutations and natural selection, it would be transformed by adapting to the conditions of its surroundings.[4]

Thus Monod associates life with the absence of a true goal, with the absence of final causes in the realization of organized forms of life, both at the level of an individual's growth from the germinal cell and from the embryo (*ontogenesis*) and at the level of the differentiation of species and their evolution toward complexity (*phylogenesis*). Note here that the alleged absence of final causes is a principle posed a priori. Nothing prohibits interpreting the paleontological and biological observations as manifesting the action of a final cause (God) at work in the very heart of the mineral world to call forth life and at the very heart of the living world to call forth spirit, as Pierre Teilhard de Chardin did several decades ago.[5] For Teilhard de Chardin, this final cause works on matter untiringly and establishes a total continuity between inanimate matter, living organisms, and thinking beings. What we call life and spirit are no more than thresholds, the crossing of which allows these immanent properties to be revealed. But the price to be paid for such a reading of the book of life is a delicate separation between the scientific field, which is that of causal laws, and the spiritual field, which is open to final laws. Here doubtless is the reason that so few scientists today commit themselves to such a path.

Formulated in a way consistent with Jacques Monod's reductionist analysis, the question of the specificity of the living comes down to that of evolution. Is it conceivable that the DNA molecules that form the genetic program of the various living species and are thus responsible for the appearance of more and more complex organisms are the mechanical product of chance and necessity?

Monod's teleonomy, which is based on the "goal" of perpetuating the species, cannot account for the complexification of life during phylogenesis. According to François Jacob, teleonomy does not limit itself to the goal of the perpetuation of the species; it is also and above all the obstinate conquest of autonomy. According to him, the living being is that which tends to live freely. Its internal program is precisely oriented toward overcoming constraints by metamorphoses that logically suppose thresholds to cross and successive births.

What is perhaps most characteristic of evolution is the tendency to flexibility in the execution of the genetic program; it is an "openness" that allows the organism constantly to extend its relations with its environment and thus to extend its range of action. In so simple an organism as a bacterium, the program is carried out with great rigidity. It is "closed" in the sense that the organism can only receive very limited information from its environment and can only react in a strictly determined way to this information. . . . At the macroscopic level, therefore, evolution depends on setting up new systems of communication, just as much within the organism as between the organism and its surroundings. At the microscopic level, this is expressed by changes in the genetic program, both qualitative and quantitative.[6]

Thus we find that the ingredients to which the study of dissipative systems has accustomed us are preliminary to a definition of the living. Like dissipative systems, living organisms are essentially open systems as defined by the thermodynamics of irreversible processes. In order to live, they must be capable of continually exchanging matter and energy with the outside. They are this way by necessity, because the development of their complexity cannot be realized except through consumption of a preexisting negentropy furnished either directly by solar light (as with photosynthesis) or indirectly through ingesting organic matter rich in energy and charged with negentropy. But life demands this supplemental dimension of teleonomy, of a goal, of the conquest of liberty. According to the synthetic theory of evolution, this orientation is imposed by natural selection. The question that thus arises can be put in these terms. Is the existence of an internal genetic program allied to the external solicitations of the environment to which life responds through the mechanical selection of the most apt sufficient to account for our paleontological and biological observations? As a convinced neo-Darwinist, Monod firmly believed this. But it seems better, here again, to admit our uncertainty on the subject.

About ten years ago a passionless examination of the known facts could not persuade an independent thinker to embrace the neo-Darwinists' faith unreservedly. Their militant attitude, doubtless comprehensible in the light of their effort to overturn centuries of religious dogmatism, was not without some degree of ideological prejudice. Serious theorists constructing computer simulations concluded that the proposed plan—according to which the complexity of life would increase at the mercy of random genetic mutations and natural selection—was impossible. These calculations were published in a more or less confidential form.[7] Some thinkers had, however, enough independence to proclaim these difficulties openly and to express their doubts as to the explicative power of neo-Darwinism.[8]

■ Evolutionary Surges

The prominence accorded to the genetic code as the essential and distinctive condition of the living does not make it easier to understand either the origin of life or its power to become more complex during the course of evolution. The dialogue of the living with the environment is also essential, but it does not seem possible to reduce it solely to the blind rules of chance and selection. To pin down the terms and conditions of this dialogue, François Jacob remarked that the properties of life go beyond the merely chemical composition of DNA. Life also has the characteristic property of being organized into successive phylogenetic and ontogenetic stages with interdependent levels of integration. During evolution, each stage had to achieve its own degree of organization before the organism could pass to a higher stage of complexity (for example, cellular life took a long time to perfect itself before multicellular beings appeared). Life climbs a ladder of "integrons."

At certain times in its history, evolution underwent abrupt accelerations contrasting with the peaceful perfectioning of previous stages. This was the case, for example, about 570 million years ago, when the main divisions of the living realm branched out almost simultaneously from the soft and relatively homogeneous marine beings that populated the Precambrian seas. It also seems to have been the case in the period witnessing the first tetrapodal vertebrates' conquest of dry land, about 380 or 400 million years ago. But no doubt the most remarkable of these surges was the development of the human prefrontal cortex, which practically doubled in volume in less than a million years, from *Homo erectus* to modern *Homo sapiens* (see fig. 12).

Thus, "the speed with which the human brain developed is still hard to understand."[9] It is during these past few years, with the proposition of temporal control of genetic expression and the first discoveries concerning regulatory genes, that we have begun to conceive of the types of rules that might steer evolution, in which mutations no longer appear as the consequences of pure chance but rather fit into a certain grammatical coherence (see chap. 17).

Figure 12. The evolution of brain size from humanity's ancestors up to the current epoch. From Robert Jastrow, *The Enchanted Loom: Mind in the Universe* (New York: Simon and Schuster, 1981), p. 139. Used by permission of Robert Jastrow.

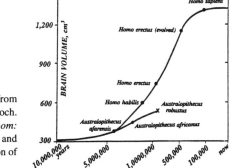

■ Which Logical Rules Does Evolution Obey?

This syntax, whose rules are not yet entirely known, shepherds chance and orients ontogenesis and phylogenesis in a well-determined direction. As applied to humankind, it inspired in Jeremy Campbell the beautiful title of his book *The Grammatical Man*.[10] He insists on the distinction between genes of structure and genes of regulation, which parallels the distinction between words and the rules of everyday grammar. By juxtaposing words not through pure chance but under the control of previously acquired rules of grammar, a young child can incessantly invent new, intelligible phrases. In the same way, over the course of evolution, nature seems to invent genetic mutation not in an absolutely blind fashion but rather by following a grammar imposed perhaps partially by the environment and partially by the regulatory genes.

On the environmental side of the equation, the growing complexity of living beings might follow the rule of "self-organization through noise," an idea originally introduced by von Foerster.[11] Henri Atlan remarked vis-à-vis this subject that the complexity of living systems was established through a subtle interplay between the internal redundancy of structures and random events. External chaos draws from the reserve of redundant structure to create new complexities. Thus, the multiple copies of some DNA sequences, frequent in many examples of genetic patrimonies, give organisms the opportunity to acquire new genes for the fabrication of proteins absent until then. Seen in this way, the genetic patrimony of the frog, which contains repetitive sequences by the tens of thousands, making it longer than the human genetic patrimony, possesses a capacity for future evolution perhaps superior to that of humankind! In addition, sexuality—which at the level of molecular biology means the systematic pairing in the same genome of two genes with the same function—is another reservoir of redundancy. Its introduction into the history of life allowed the unobtrusive appearance of recessive genes, that is, genes not leading to the appearance of new characteristics as long as they are masked by homologous dominant genes. Certain recent theories that must be noted in passing indicate that genomic redundancy could also play an important role in the process of aging and death, to the extent that it introduces a supplemental fragility of the genome. According to B. L. Strehler, the disintegration of genetic DNA through the loss of copies in the genes responsible for the production of RNA ribosomes could play an essential role in the processes of senescence.[12]

As to the grammatical rules contained in the regulatory genes, or hidden in the imposing mass of "introns" (parts of the genetic code that are not transcribed in messenger RNA), they appear to orchestrate the temporal aspect of development and to preside over cellular differentiation.

The property of autocomplexification attributed to living organisms suggests that they might have their own capacity for decision: life would have no other goal than that of organizing itself. As demonstrated in 1962 by W. R. Ashby, a closed system cannot find within itself the criteria it must apply to decide whether the organization just attained is good or bad from the point of view of increasing complexity.[13] Henri Atlan extends this argument to insist on the role of the environment.[14] The apparent goal of life is dictated from the outside, and specifically through random fluctuations of the environment to which life is submitted. It seems paradoxical that a goal, or even a semblance of one, would emerge from the accidents of the environment, but this fits with my analysis of the role of cosmic chance in the arrow of physical time. It must be recognized, however, that we are still far from having a complete understanding of these mechanisms.

■ The Origin of Life

When we know everything there is to know about the grammatical rules presiding over the architecture of life, we will perhaps be able to understand the mechanisms by which life appeared on the Earth. Many think that the exceptional conditions reigning on the surface of the Earth 3 or 4 billion years ago favored the passage from the inanimate to the animate. At the University of Chicago, in 1953, the experiments of Stanley Miller and Harold Urey on the production of amino acids in a gas supposed to represent the primitive atmosphere and subjected to powerful electrical discharges (simulating the then numerous and violent storms) gave hope of understanding the beginnings of life, for certain simple organic molecules (amino acids) appeared in the inert mixture. But the terrestrial atmosphere of the age, we know now, was poor in methane and rich in carbonic gas, making this type of scenario unlikely. A bit later Francis Crick and Fred Hoyle formed the hypothesis that life might not have been born on the Earth but brought here from the outside, by some type of germ contained in meteorites. This hypothesis was based on the observation of some simple organic molecules found in an analysis of the absorption spectra of interstellar dust, as well as in the micrographic structure of some meteorites that resembled the skeletons of living microorganisms. In a recent article Antoine Danchin,[15] while still advancing the reasonable hypothesis that life could have arisen from a mineral universe, recognized that the problem of the origin of life "remains as open as ever." All the scenarios in which the first autoreplicating molecules might have appeared by pure chance, either on Earth or in some cosmic milieu, are unsatisfactory. Most of the other organic molecules formed in the same conditions would represent poisons for the mechanisms of replication of life and would surely have killed life from the very earliest stages. . . . The first living molecules belonged perhaps to the class of nucleic acids (an idea proposed first by C. Woese and L. Orgel)[16] rather than to the classes of the amino acids and proteins. This hypothesis is supported by the observation of a bit of ribosomic RNA of the species *Tetrahymena pyriformis,* an archaic protozoan, which already possesses some catalytic properties thought until now to be reserved to enzymes. Must the origin of life be linked to nucleic acids or to proteins? The debate, as Danchin noted, strongly resembles the old question as to which came first, the chicken or the egg. The answer is now leaning more toward nucleic acids, but nothing has yet been definitively settled.

■ ■ ■

This quick look at nature and the origin of life proves once more that the great questions still remain unanswered, although their terminology has been refined. Current thinking on the nature of life returns persistently to the ancient questions of Aristotle, who was already insisting, twenty centuries ago, on the importance of organization as a distinctive criterion of the living, going so far as to identify *the soul* as animating inert matter and giving it life by *organizing* it at the most basic level. As Michel Guédès remarked, such a thesis is remarkably close to the modern understanding when the Aristotelian soul is replaced by information.[17]

But organization is constructed step by step over time. Time thus plays a fundamental and active role in the evolution of species and the development of individuals. In this sense, no operational definition of the living can be given if it does not include a truly temporal dimension. Such a dimension was not explicitly present in the writings of Jacques Monod, for example, and it is perhaps this that makes his reasoning fragile. More than the mineral world, and perhaps in a qualitatively different fashion, the living world finds itself under the sway of time. This is a world of vibrations like that of the atoms, but where the vibrations determine not only levels of energy but the form itself, or more exactly, the manner of existence of the subjects considered. It is a world, moreover, where the vibrations weave an inextricable network of interconnections, of cooperation, of resonances with the rhythms of neighboring units of life or with outside rhythms. The moment has come to assert not only that time plays a primordial role in the construction of life but also that its rhythms are necessary to the maintenance of life in its ordinary expression and in its daily functions.

Life: A Bundle of Intertwined Cycles

The rigorous conditions necessary for the emergence and maintenance of life render it a sort of "continuous miracle," which is made possible by the appearance and refinement of all kinds of cybernetic mechanisms controlling the conditions of the internal environment. Life is marked by a constant effort to thwart variations in the outside environment and to regulate conditions in the interior environment. By way of these mechanisms of regulation, the living being progressively won the autonomy and the longevity necessary to survive the periodic upheavals of nature. It learned to survive first the alternation of days and nights and then the unforeseeable catastrophes such as droughts or the abrupt disappearance of a food source. These observations were not understood, however, until quite recently. The concept of homeostasis was introduced in biology by Walter Cannon in 1926. Science has not yet acquired the necessary perspective on this recent concept to see that homeostasis is most often only a global view of a very complex network of cycles that manifest themselves in rhythmic oscillations of most vital parameters.

■ From Homeostasis to Chronobiology

Homeostasis is the principle by which the living organism regulates its internal environment so as to keep constant its principle factors—temperature, concentration of the different chemical substances, and so on. But this impression of constancy in the internal environment is true only on a superficial level. Examine the case of a small portion of intestine. At first sight, its aspect remains the same throughout the hours and the days. A more attentive observation reveals an incessant labor taking place beneath the surface of this apparent sameness. The cells that line the inner wall of the organ, which presents numerous and deep villi, are continually replaced.

The cells situated at the summits of these villi disappear through natural attrition or are broken loose by the transit of foods through the intestine. They are replaced by new cells resulting from cellular divisions in the hidden depths that then migrate toward the summits of the villi. Replacement during the life of an adult is continuous, total, and relatively rapid. The cells at the bottom divide on the average once per day.

Alain Reinberg, a pioneer in this field of research, which is called chronobiology, tells in one of his first books of his surprise when he and his collaborators discovered the cyclic variations of the amount of salts in urine, and especially those of the potassium ion. "Our surprise at this finding was great. No one, during our university studies, either at the Faculté des Sciences or at the Faculté de Médecine, had ever spoken to us of these rhythmic variations. On the contrary, the teaching I had received led me to think that an entire ensemble of neuro-endocrine systems functions in the organism in such a way as to keep the cellular and extracellular environments constant."[1]

A few decades ago it was rediscovered that homeostasis is only an overall aspect of the situation and that at the local level many parameters undergo rhythmic variations. Living beings thus depend on a very complex network of cycles evidenced by regular oscillations. Glycolysis and mitosis of the cells of the intestinal villi are but two examples among many.

Many of these rhythms have a period close to twenty-four hours. For this reason they are called circadian cycles. Others have rhythms close to a year. They are called circa-annual cycles. Finally, others have intermediate rhythms, such as a woman's fertility cycle, which is close to the period of the lunar revolution.

Despite their simplicity, the primitive unicellular procaryote beings (without a nucleus) already show evidence of circadian rhythms, although the literature on this subject is not yet abundant and the observations should be confirmed. It should be noted that these primitive species have at least one precise rhythmic activity: that of their reproduction, which is accomplished by simple division at more or less regular intervals.

With protists (unicellular beings that possess nuclei and show great internal complexity), certain cadenced activities corresponding to the day-night alternation can be clearly distinguished. The case of the diflagella of the species *Gonyaulax polyhedra* has been well studied.[2] These microscopic organisms are responsible for the phosphorescence of marine plankton. They demonstrate a remarkably cyclic photoluminescent activity with a period of twenty-four hours. This activity persists when the animals are placed in a continually lighted surrounding, although the intensity of the phenomenon rapidly decreases when the level of the ambient luminosity rises. The period of photoluminescence of these protists is thus not strictly linked to the alternation of days and nights. In a constant luminous ambience, the value of the period departs slightly from twenty-four hours. A colony of these animals subjected to an abruptly modified day-night alternation continues for a time to follow its previous rhythm before adapting itself to the new periodicity.

The photosynthetic activity of plants is also patterned on the day-night alternation. In lighted surroundings, chloroplasts capture photons and use their energy to separate molecules of water into oxygen, which they release into the atmosphere, and hydrogen, which they keep. These hydrogen atoms are then combined with the atmospheric carbonic gas to synthesize the carbohydrates of which the plants are constituted. Other types of circadian rhythms in plants have also been observed.[3]

Mitosis, or reproduction of cells through the duplication of chromosomes, also takes place in accordance with an immutable and precisely regulated ritual. The duplication of the DNA contained in the nucleus of the cell begins after a phase of normal cellular activity of variable duration. The cell nucleus swells slightly, which is a precursor to the beginning of reproduction itself. Reproduction takes place in several phases: prophase, prometaphase, metaphase (during which the chromosomes become more and more clear and organize themselves), anaphase (during which the chromosomes separate into two groups and migrate), telophase (during which each group of chromosomes re-forms a separate nucleus), and cytoplasmic division. The entire process takes about an hour (for a plant) or a day (for a human). The rate of cellular renewal, which is not to be confused with the duration of cellular division, is extremely variable according to the tissue. The cells of the deep dermis divide once every four days, red blood cells are replaced every four months, and so on.

■ Hormonal Clocks

There are rhythmic biological functions other than those linked to reproduction and to the intracellular exchanges of energy. Those linked to the transport of active substances or of information between organs are also regulated by rhythms, many of which are circadian. Hormones produced by the endocrine glands and transported toward their target by the blood play a well-known role in the human body. Some of the most important are produced by the pituitary gland, which is situated at the base of the brain and controlled by neural networks. The pituitary gland produces in particular ACTH, or the corticotropic hormone, in a circadian mode with a maximum marked in humans by dawn (and a minimum at the fall of night). ACTH controls, in part, the activity of the adrenal glands, which are charged in their turn with producing the corticoids, whose importance in the economy of the human body is known. The protective action of cortisone in some allergic ailments such as asthma explains why the more frequent and the most grave crises generally occur during the night, when the amount of this hormone in the blood is at its lowest. These nocturnal crises are not linked only to psychological distress, as was previously thought, but have a certain physiological basis.

■ Oscillations in the Central Nervous System

The pituitary gland's production of hormones is one form of message addressed by the brain to the different organs of the human body. But the principal means of interorganic communication in the body is the nervous system. The evidence shows that the nervous system is also a site of oscillations and rhythms. The individual nerve cell is already, in certain respects, an oscillator. Let us examine the constitution of a typical nerve cell. Its cellular body is extended by a filament, or "axon," which is specialized in the long-distance[4] transport of nerve impulses. The axonic wall is endowed with multiple microscopic channels that allow (according to their nature) sodium or potassium ions to pass. Their opening is governed by the membrane potential.[5] Once the potential crosses a certain threshold, due to the influxes received from the cellular body or the dendrites,[6] the sodium channels open, engulfing a tidal wave of sodium ions swarming into the cell. After a short delay, the potassium channels open in their turn, and the corresponding ions escape from the cell. This ebb and flow of ions determines an action potential, a violent and transitory variation of the potential of the membrane that goes locally, in less than a millisecond, from -70mV to $+30$mV before progressively returning to its initial value. This electrical shock travels along the axonic fiber, constituting the active and significant part of the nerve influx. But the cell membrane contains other channels with much slower time constants. Some, permeable to potassium ions, allow these ions to escape into the external environment. Others, permeable to calcium, allow these ions to penetrate into the cell. The play of these slow channels determines oscillations without propagation of the membranous potential, with a period that can be as long as ten seconds. But if, during these oscillations, the membrane potential crosses the threshold value characteristic of the opening of the sodium channels, an action potential is generated and propagates itself the full length of the axon. These action potentials follow each other regularly, appearing one after the other or in bursts separated by the period of the original oscillations. Such regular cycles are frequently observed in recordings made using a microelectrode introduced into or in the neighborhood of a particular neuron. One well-known example is the R15 neuron in the abdominal ganglion of the *Aply-*

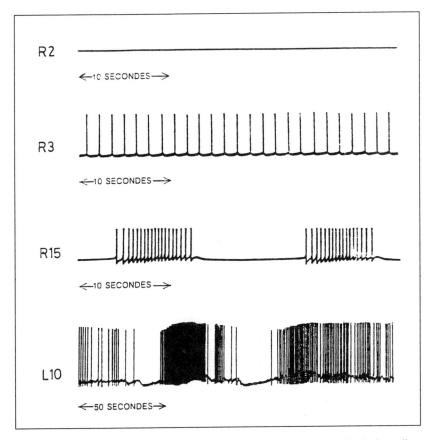

Figure 13. Different modes of discharge of some neurons of the *Aplysia's* abdominal ganglion. From Eric Kandel, "Small Systems of Neurons," *Scientific American*, Sept. 1979. Used by permission of Eric Kandel.

sia, a sea slug whose neurons are few and large enough to make it a choice subject for neurobiological research. The cyclic activity of this neuron, already visible when its activity is recorded in a living and whole animal, persists when the nerve ganglion is isolated and even when the neuron itself is isolated. It is therefore an autogenous rhythm (fig. 13).

■ Astrology and Chronobiology

The majority of the internal rhythms studied by chronobiology are circadian rhythms, that is, they follow the natural succession of days and nights. Other rhythms are circa-annual; still others, such as that of feminine fertility, pose the question of their links to the 29½-day lunar cycle. There is therefore a strong presumption that the periods of these cycles are in accord with astronomical cycles. Do the stars govern life and its pulses, as the ancients believed?

A closer examination indicates that, in general, biological rhythms do not rigorously conform to astronomical rhythms. This can be seen by examining their periodicity in surroundings free from astronomical constraints. The example of the luminescence periodicity of the *Gonyaulax polyhedra* demonstrates this. In a constant artificial light it varies from twenty-four hours. Another rhythm, primordial for humankind, is that of the successive states of waking and sleeping. As seen with subjects forced to live without a clock and without contact with the outside, the rhythms of sleep, of body temperature, and so on, spontaneously take on a value different from twenty-four hours, generally in the neighborhood of twenty-five hours. This difference between the astronomical rhythm and the endogenous rhythm explains why we tolerate jet lag in westward transcontinental flights better than in those heading toward the east. An organism traveling toward the west does not feel an hour of jet lag because it brings the hour of sleep closer to the moment fixed by the endogenous clock. On the other hand, an hour of jet lag in eastward flights leads the organism to a supplemental discrepancy with respect to its endogenous rhythm.

Curiously, it seems that the waking-sleeping rhythm in *Homo sapiens* when placed in conditions of artificial light stays at the value of twenty-five hours only for a certain time. When the period of total isolation is prolonged—in a cave, for example—it has been noted that after about two weeks the sleep cycle passes spontaneously "to the higher fourth," that is, to about 33.4 hours, or, conversely, plunges to half this value. The rhythm of temperature, however, persists with a period of twenty-five hours. The organism thus spontaneously chooses an arrangement of its own endogenous rhythms different from that which it adopts in natural light.

Our endogenous rhythms are thus *synchronized with,* and not *controlled by,* astronomical rhythms. They must thus have their own internal control clocks. Cosmic rhythms influence living beings, but only to force the natural cadences to align with them and, doubtless, to select the most viable forms of life.

Indeed, it is easy to imagine that life could organize itself spontaneously according to the cycles or the rhythms of varied periods but that only the species having their internal clocks more or less synchronized with astronomical rhythms have survived and have been selected over the course of evolution. They possess, in fact, appreciable advantages in the struggle for life. Do not circa-annual rhythms allow the most profit to be drawn from the alternation of the seasons and from its effect on the abundance or the scarcity of food? In the same way, a nycthemeral rhythm of waking and sleep permits certain species to maximize their activity when the conditions of visibility are best and allows other species to take advantage of their prey's nocturnal rest to better surprise them. Finally, it is indispensable that the cycles of photosynthesis be in accord with or adapt themselves to the natural rhythms of light. Any other solution, if it had been envisioned by nature over the course of evolution, was rapidly eliminated.

It is not quite clear what genetic advantage could encourage the feminine endocrinal period to correspond to the lunar cycle. The feminine endocrinal rhythms of other mammals are extremely varied, and in the human species their period varies greatly from one woman to another. The approximate coincidence of this cycle with the lunar cycle is perhaps fortuitous. Lovers of moonlight will doubtless prefer to continue to think the contrary, and perhaps they are right!

■ Toward the Localization of Internal Clocks

But where are the control clocks of the organism's internal rhythms, and how do they function? The vinegar fly (*Drosophila*), much used since Thomas Morgan for all studies of genetics, also furnishes a preferred model for the understanding of certain endogenous timekeepers.

This insect's natural circadian rhythm of larval hatching has been thoroughly studied. When they are raised under constant lighting, *Drosophila* larvae metamorphose into nymphs (pupae) that hatch at any hour whatsoever, seemingly at random. The hatchings are produced in a continuous and incoherent fashion over a long period of time. But if at a given moment the larvae are plunged all together into total darkness, the pupae hatch in widely separated batches, with an interval very close to twenty-four hours. Thus, the internal clocks of all the larvae seem to have been "reset" by this procedure at the exact instant they were transferred into darkness. In experiments of this type, everything happens as if the larvae's conditions of maturity were continually compared to an internal clock that runs through its cycle and gives the insects "permission to leave." Figure 14 assembles the results of more than a hundred different experiments totaling about forty thousand hatchings.

There is good reason to believe that the internal clock regulating *Drosophila* hatching is located in the insects' brains. In fact, the photosensitive receptors that determine synchronization with the diurnal rhythm are located in this region of the body, and the deregulations sometimes observed are always linked to mutations affecting cerebral tissue. A spectacular example of such a deregulation is furnished with a specific mutation in *Drosophila melanogaster* that gives birth to flies whose internal clocks are reg-

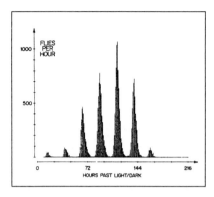

Figure 14. Rhythmic hatching (with a period approaching twenty-four hours) of the *Drosophila* pupae, placed in darkness at time $t = 0$. From A. T. Winfree, *The Geometry of Biological Time* (New York: Springer-Verlag, 1980), p. 404. Used by permission of Arthur T. Winfree.

ulated on a period of nineteen hours rather than the usual twenty-four. The corresponding stock was the subject of systematic breedings. All the descendants possess a clock whose basic period is nineteen hours.

This observation confirms the nervous origin of the clock regulating *Drosophila* hatching on a circadian mode. But it also gives evidence of its deep-seated origins, since a genetic mutation is capable of affecting it. Thus, it is distinctly possible that the clocks regulating the cycles of behavior or function depend on the expression of certain genes. Their mechanisms would be inscribed in the book of the genetic patrimony.

In the specific case of the *Drosophila,* this hypothesis recently received an astounding confirmation. The gene regulating the temporal behaviors of the *Drosophila* (including the rhythm of its hatchings, its circadian cycle of activity and rest, and the periodicity of the males' songs) was localized and identified. The new techniques of genetic manipulation allowed an artificial mutation to be induced in this gene.[7] This mutation modifies these behaviors and the periods of their rhythms. The study also confirms the hypothesis that the master clock regulating the periodic behaviors of this animal are located in the brain.

No doubt similar clocks exist in our brains and regulate the cadences of our daily behavior. The circadian cycles of the hormones are under the control of the brain's endocrinal glands, notably the pituitary gland. This latter is linked directly to the hypothalamus, which is both an integral part of the central nervous system and an important endocrine gland. It is now thought that a master clock located in the suprachiasmatic nucleus of the hypothalamus controls the hormonal secretions of the pituitary gland, alimentary behavior, the succession of waking and of rest, and so on.[8] It would, however, be an error to conclude that there is one single cardinal clock responsible for all circadian cycles. The organs themselves obey their own circadian cycles (this is true, for example, of the adrenal glands). The production of serotonin in the pineal gland drops sharply at night, but this gland is known to be under the control of external stimulations through visual pathways. As do those of other fundamental periods, the circadian clocks seem numerous and widely distributed in the organism. But they are coordinated.

■ ■ ■

The diversity and number of cycles regulating the life of a living organism do not allow us to pinpoint a particular master clock for each function. There is often synergy between the different vital oscillators, and the maintenance of these synergies doubtless represents one of the principal conditions of the maintenance of life. Thus, thanks to the works of the German psychiatrist Hans Berger in 1929, we know that cerebral activity is often translated at the level of the scalp not in disordered fluctuations of electrical potential but in organized waves clearly visible on an electroencephalogram. The "alpha" rhythm that is observed to coincide with the state of calm wakefulness, where attention is not solicited in any particular way, is characterized by widely spread oscillations of the electrical potential with a frequency close to 10 Hz (10 oscillations per second). Researchers must look for the origin of these waves in the synchronization of millions and millions of neurons in the deepest levels of the brain, probably at the level of the reticular formation. Waves of this type might play an important role in successive perceptions and in their linkage.

We thus can see that time participates closely in cerebral activity, that it commands hormonal cycles, imposes organic rhythms, and finally, determines the cadences of behavior. It impregnates all the tissue of the living individual; it participates in the individual's maintenance and in the coordination of its functions. But time presides also over the individual's development, in the progressive elaboration, from the original egg, of an adult being belonging to a specific species, with its own morphological characteristics.

18

The Temporal Control of Development

The relatively simple architecture of the vinegar fly's genetic code and the rapidity of its reproduction make it a choice model for studies of genetics and morphogenesis (the development of the embryo). The discovery of the gene "per" (period) in the *Drosophila* and mutations affecting it has contributed greatly to demonstrating that the rhythmic character of the hatchings is written into this animal's genetic heritage. But does not time have other important functions, other relations with the genetic code and its expression in an individual? In fact, not only do the embryos of living beings develop *in* time, but time itself, in a way, directs their final destinies. Before addressing this question, it would be a good idea first to describe or to review briefly the essence of the rules of this code.

■ The Fundamental Principles of the Genetic Program

By 1950 desoxyribonucleic acid (DNA) was already known to exist in the nuclei of cells. It was also known to contain repeated assemblages of four small nitrogenous molecules: adenine (A), thymine (T), guanine, (G) and cytosine (C). But it was not yet known that the precise sequence of these bases in DNA formed the genetic heritage. Studies of X-ray diffraction of DNA molecules of certain viruses, conducted notably by Rosalind Franklin and Maurice Wilkins in London, and the essential observation made that same year by Erwin Chargaff of the universal equality of their content in adenine and thymine, on the one hand, and in guanine and cytosine, on the other, allowed James Watson and Francis Crick to elucidate in 1953 this molecule's stereochemical structure. It also allowed them to explain the mechanisms by which DNA is replicated (during cellular division) and used for the fabrication of proteins.

DNA consists of a double helix, each strand being formed by a chain of molecules of phosphoric acid and desoxyribose. This exterior scaffolding gives DNA an excellent resistance (it is preserved, for example, in the cells of Egyptian mummies) while leaving it enough flexibility to allow it to coil around on itself in the chromosomes.[1] To each desoxyribose radical is also attached, toward the interior of the helix, one of the four bases A, T, G, or C. Each of these bases is paired with the base that faces it, attached to the carbohydrate radicals of the other helix.[2] An excellent large-scale image of DNA's microscopic structure is furnished by the beautiful double spiral stone staircase of the castle of Chambord. In DNA the exterior scaffolding is constructed of desoxyribose and phosphoric acid, and the steps of the staircase are constituted by the basic couples A-T and G-C.

When cellular replication takes place, the DNA twists are released and the two strands are separated through the action of specific enzymes. The opening of the two strands allows the sequence of the adjoining bases of each strand to appear, like a sequence of letters in an alphabet with four signs. This sequence forms a word or a phrase that contains the recipe for fabricating the protein coded in the gene under consideration.

■ From DNA to Proteins

We also know how this fabrication takes place concretely. There is first a transcription phase. After the helix opens and its strands separate, the sequence of bases coding for a protein (its "gene") is exposed to the external environment. A complementary base glues itself to each of the bases in such a way that there is a progressive formation of a "negative" copy of the original gene; the corresponding molecule, or messenger ribonucleic acid (m-RNA), detaches itself from its support and emigrates toward the "ribosomes," which are subsidiary factories scattered throughout the cell. Then the translation phase begins. The ribosomes receive not only the messenger RNA but also other specialized molecules of nucleic acids (the transfer RNA) to which are attached the molecules of amino acids, the bricks of which proteins are truly constructed. In the ribosomes, each group of three nitrogenous bases presented by the m-RNA provokes a particular amino acid to attach to the soon-to-be-born protein. Thus, each group of three letters in the nucleic alphabet is "translated" into one letter in the proteinaceous alphabet, which is constituted of the twenty different types of amino acids.

Incidentally, a key element of chronogenetics must be noted. For a gene to express itself and a proteinaceous molecule to be fabricated, it is first necessary that the corresponding portion of DNA be untwisted and that the two strands of DNA be separated. These operations are under the control of specific enzymes, themselves synthesized from genetic instructions. It is as if the way to read a large book meant to be read nonsequentially were to be found written in some part of the work—in the introduction, for example. The openings and closings of the different chapters are directed by an ensemble of regulatory genes whose strict temporal orchestration is essential.

■ The Genes of Development

The *Drosophila* model's great success in establishing the basis of modern genetics at the beginning of this century, particularly as concerns genetic mutations, has not been shaken by the recent deeper exploration of the genetic code in its temporal dimension, in particular, its role in cellular differentiation and morphogenesis.

Thanks to these studies, we now know that organisms possess genes of development that control, inhibit, or activate the expression of other genes fabricating structural or other proteins at different stages of development. In the embryo this temporal control of genetic expression defines the lineages; that is, it directs the cells produced toward the formation of specific organ tissues. These genes of development allow the expression and subsequent repression of other genes at certain well-determined periods of ontogenesis.

Three types of developmental genes can be distinguished in insects: genes of maternal effect, genes of segmentation, and finally, homeotic genes, which will be more specifically examined here.

The genes of maternal effect, which are present in the egg before fertilization, determine the egg's general orientation, its "polarity." The role of these genes was proven by the observation of mutations leading to monsters possessing two-joined heads and without the lower half of the body, or two abdomens without a head. These mutations are to be observed when the polarity is destroyed or when the expression of the genes that determine it is impeded at the beginning stages of development.

Operating at the larval stage, the genes of segmentation determine the number of segments these animals will exhibit. The *Drosophila*, for example, comprises the mandibular, maxillar, and labial segments, which form the head; the prothorax, mesothorax, and metathorax, which constitute the thorax; and finally, the eight abdominal segments. Some mutations affecting the genes of segmentation lead to the suppression of certain segments.

Finally, the third category of developmental genes consists of the homeotic genes, which determine the nature of these segments. These genes are named "homeotic" because the mutations affecting them result in the appearance in a segment of inappropriate organs or tissues that are normally specific to other segments (homo-eo). The mutations in these genes have

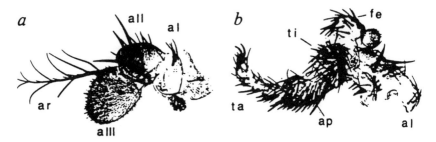

Figure 15. The activation of two homeotic *Antp+* genes in the mutants of the *Drosophila* fly carrying this duplicate gene provokes the appearance of legs (b) in the place of antennae (a). Following S. Schneuwly, R. Klemenz, and W. J. Gehring, "Redesigning the Body Plan of *Drosophila* by Ectopic Expression of the Homeotic Gene *Antennapedia*," *Nature* 325 (1987): 18. © 1987 by Macmillan Magazines Ltd. Used by permission of Walter J. Gehring and Macmillan Magazines Ltd.

spectacular consequences, for they give rise to monsters possessing, for example, supernumerary legs where antennae usually grow.

Two homeotic genetic complexes have been especially well studied in the *Drosophila*: the Antennapedia complex (ANT-C) and the Bithorax complex (BX-C). These complexes are located on the right arm of the insect's third chromosome. The mutations of these genes determine an abnormal differentiation of the head or the anterior segments (if it is a question of mutations affecting ANT-C) or of the thoracic or abdominal segments (if they affect the BX-C complex). It has been possible, using the techniques of genetic manipulation, to produce mutants that possess two *Antp+* genes with a particular promoter[3] that can be activated by a thermal shock. If the promoter has not been activated, the larvae possessing this genome mature into flies that are normal in the adult state. If a thermal shock is applied at an early stage in the larval development, however, thus activating the promoter, then the proteins corresponding to the gene *Antp* are produced in a superabundant fashion. The adult animals will then carry feet in the place of antennae—the appendage characteristic of a posterior segment will have replaced that which characterizes an anterior segment (fig. 15).

The homeotic genes of the *Drosophila*, then, do determine the architecture of the adult animal. We now know that a single mutation in genes of this type can induce an extensive reorganization of the body. Of course, we had already seen mutants of certain mollusks whose shells spiraled in the direction opposite the species' norm. The existence of a mutation in humans called *situs inversus*, which induces a complete reversal of right-left symmetry, has been reported: the heart is located on the right and the liver on the left, the intestines coil in a reverse direction, and so on. Humans, however, do not exhibit major organic mutations (such as the appearance of a leg in the place of an arm) comparable to those affecting the homeotic genes in the vinegar fly. Perhaps, during embryonic life, the map for the organization of superior organisms does not follow the scheme of segmentation seen in insects.

■ Structure of the Homeotic Genes

The molecular map of the *Drosophila*'s homeotic genes shows that they are made up of short sequences of DNA (key fragments) found in identical or nearly identical form in several regions of the ANT-C and BX-C complexes. These fragments, including about 180 nuclear bases, contain the fabrication code for a protein and its sixty amino acids. They have been called "homeoboxes." They are the spatio-temporal regulators of genetic expression, even though the mechanism by which they function is not yet well understood. They are thought to direct the synthesis of proteins that, in combining with certain parts of DNA, determine or impede the execution of these fragments' corresponding programs according to the general control plan of genetic expression discovered by Jacques Monod and since known by the name of *operons.*

Suspecting that homeoboxes could be key fragments in the temporal control of genetic expression, Walter Gehring and his collaborators[4] sought to determine whether other species might not include sequences of homeoboxes identical or closely related to those discovered in the *Drosophila.* They thus began to look for them first in other insects and in certain worms considered to be ancestors of insects and then in certain vertebrates. To their great surprise, the researchers discovered that batrachians possess homeoboxes incredibly similar to those of *Drosophila.* Fifty-nine of the sixty amino acids belonging to the protein coded by these homeoboxes are identical and at the same place in the genes of the two species! They also discovered comparable genes in mice and finally in humans. In 1988 the complete analysis of the twenty-nine different homeoboxes had been finished, of which ten had been found in the *Drosophila* and nineteen in other species. The homeobox of the *Antp* gene in *Drosophila* and that designated as *C1* in humans have an 86 percent degree of homology for the sequences of the nucleic bases and 98 percent (fifty-nine of sixty) for the associated protein.[5] This similarity suggests that homeoboxes play a role sufficiently important that they have been preserved almost without change throughout evolution. All the homeoboxes of which I have so far spoken concern species with segmented bodies in homologous parts (in vertebrates, these segments may correspond to vertebrae), but a homeobox has recently been discovered in a species of sea urchin, an organism that does not appear to have any segmentation. Thus, it is possible that the homeoboxes have, in the temporal control of genetic and morphogenetic expression, a more general role than just determining the fate of different body segments.

Although all the required evidence has not yet been assembled, it seems that homeoboxes effectively play a role in controlling the organism's development, in particular, in specifying its architectural plan. The discovery of such temporal regulations is of the utmost importance. It will surely enable us to understand how, with an almost identical basic genetic patrimony and embryos whose resemblance to each other is remarkable, different species confront such different destinies. Through the play of inhibitions and activations of genes, the temporal program dictates to the cell which products to fabricate, which organelles to construct, and which specializations to acquire. It thus orients the differentiation of cells and the characteristics of species whose basic embryonic materials seem to have, at their origins, a great polyvalence.

■ The Ubiquitous Resemblance of Embryos

The absence of initial specificity in the living cell had already been remarked on, in the last century, by the inventor of the theory of recapitulation, Ernst Haeckel. Figure 16 reproduces the sketch he published in 1874, which compares the morphological aspect of embryos of species as diverse as the chicken, the dog, the tortoise, and the human at intermediary stages of their development.

Thus, from the original clay of the egg, the temporal program alone commands the different switches and paths, guiding the embryo toward the species' definitive and specific form. Not only is this rule true for inferior species; it seems also to apply to humans and the higher mammals. In humans, the proteins fabricated by diverse individuals are basically the same, with only minor variations. The variability observed is nearly the same between individuals of different races as between individuals of the same race. This observation is the basis for most contemporary biologists' opposition to even the idea of biological racism.[6] Moreover, 99 percent of the proteins fabricated by the chimpanzee or the gorilla are the same as the proteins fabricated by human cells. More precisely, on a sample of forty-four homologous proteins analyzed by American researchers, the differences in formula observed in the sequences of amino acids applied to less than 1 percent of them. It is true that a small difference in the chemical formula of a protein can lead to large variations in its action if this difference is found in a key portion of the macromolecule. However, for most of the analyzed proteins, almost no difference in their specific chemical activity is observed. The alpha and beta chains of hemoglobin, for example, are absolutely identical in these primates and in man.[7]

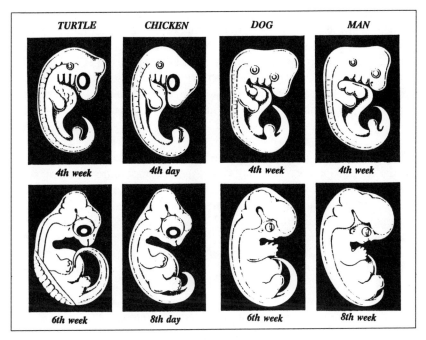

TURTLE	CHICKEN	DOG	MAN
4th week	4th day	4th week	4th week
6th week	8th day	6th week	8th week

Figure 16. Comparison of the embryos of four different species of vertebrates at different stages of their development. Above: tortoise (fourth week), chicken (fourth day), dog (fourth week), human (fourth week). Below, the same embryos at a slightly more advanced stage: tortoise (sixth week), chicken (eighth day), dog (sixth week) and human (eighth week). From E. Haeckel, *The History of Creation; or, The Development of the Earth and Its Inhabitants by the Actions of Natural Causes* (New York: D. Appleton, 1876), plates 2 and 3.

■ The Diversity of Life

All these observations lead us to conclude that humankind does not distinguish itself from the other primates principally by the information contained in its genetic code or the chemical composition of cellular constituents. Our species distinguishes itself through the temporal program regulating the expression of the genetic code, which specifically allows the remarkable complexification of the human central nervous system at a crucial moment of human development.

We think, therefore, that the raw contents of the genetic code, or the number of its significant letters, are not enough to determine the degree of complexity of the organism to be built. The example of the frog, whose genetic book is one hundred times longer than the human's, is enough to make the point. There are, in fact, in the DNA of frogs and in that of humans (but even more in frogs) innumerable repetitions and long mute sequences that considerably elongate the genome. Jean-Pierre Changeux justly asks the following question: how can a genetic code that contains perhaps a million original significant letters code for the thousands of billions of synapses (the contacts between nervous cells) scattered throughout the human body?[8] Did the molecular biologists of the 1960s or 1970s overestimate the power of the written letters? Did they, perhaps, neglect a certain dimension of the living, what one might be tempted to call the fourth dimension, even though it does not amount to physical time?

■ ■ ■

The preceding chapter suggested the power time has over life in terms of all the rhythms and cycles whose number and importance I did no more than scan. To that picture has just been added the power that time exercises on ontogenesis and in the differentiation of species. In all these cases, time manifests itself in the living being in the form of a multitude of internal clocks that govern both the being's destiny and its survival. It remains to be shown how the coordination of an organism, in its development as well as in its everyday functioning, depends on the coordination of these multiple clocks. On this question, biological research finds itself without any definitive answer. The synchronization of innumerable clocks with varied spontaneous periods, whose existence throughout the entire human body we can only suspect, presupposes mechanisms of control that should be extraordinarily detailed. It constitutes, perhaps, the keystone itself of life, insofar as this *conspiracy,* and it alone, would allow the preservation of life in individuals, if not even of groups. Are not many cancers due to a deregulation of the delicate mechanism of temporal control of genetic expression, a deregulation that allows certain types of cells to proliferate anarchically at times other than those foreseen in the map of embryonic development? The corresponding genes, whose deregulated temporal expression determines the appearance of the malignant proliferation, have been named oncogenes. Their study has developed spectacularly these past few years, to the point that it has been written that "many, perhaps all the human cancers, are linked from the start to genetic mutations."[9] Death itself will perhaps one day be explained more lucidly and convincingly in terms of the dysfunction of these clocks or this synchronization rather than in terms of the accumulation of genetic errors, lysosomic wastes, and cellular degeneration.

The Brain, the Mind, and Time

As do all organs, the brain consumes oxygen and sugar. Like them, it is the seat of rhythmic activities. But the brain is also a "logical" organ similar to the logical unit of a computer. Its job consists of "calculating," that is, receiving and then transforming sensory information into actions. This function obviously requires new mechanisms of temporal coordination. Finally, the brain is considered to be the material support of thought. The awareness of lived duration characteristic of thought might find itself shaped by temporal neuronal mechanisms of a particular type. The relationship between the time of the brain, considered in its physiological and functional aspects, and the time of thought, construed as the will's conscious projection and anticipation of the envisioned acts, is a legitimate subject of inquiry. The neurosciences attempt to describe the mechanisms that generate these particular rhythms, to study their articulations and their organization into networks, so as one day to answer questions about their relationship with psychological time.

Modesty is imperative here. The neurosciences are young, and they are still evolving. Their late arrival on the scene of the sciences can be explained in part by the prudence that the subject inspired in researchers of the past centuries for philosophical or religious reasons. More recently, it can be explained also by the overpowering influence of computer technologies, whose successes—and the excessive hopes that a certain part of society placed in them—have served to mask the original nature of biological thinking systems. But today the neurosciences are progressing rapidly. The answers they are giving are often no more than hypotheses, sometimes supported by observations that are still fragmentary.

■ The Development of the Neurosciences

The first in-depth studies on the brain were the work of the great histologists and physiologists working at the end of the last century. The painstaking microscopic observations by Ramon y Cajal showed that the nervous system consists of individual cells separated one from another. It does not form a continuous network, as had been previously thought. The term *neuron,* which designates these individual cells, was adopted in 1891 (Wilhem Waldeyer), and that of *synapse,* designating the regions of contact between two consecutive neurons involved in the transmission of the nervous impulse, was proposed in 1897 (Charles Sherrington). The majority of the guiding principles behind today's neurosciences, however, were born after World War II: the plasticity of the synapses as the explanatory element in the mechanisms of memory (Donald Hebb, 1949); the mechanisms by which action potentials are propagated in the nerve fibers (Alan Hodgkin and Andrew Huxley, 1952); the existence of neurotransmitters, which are molecules that nerve cell endings subjected to action potentials release in the intersynaptic space and which are responsible for transmitting nerve impulses (known since 1941); the existence of inhibitory synapses, which do not transmit the nerve impulse from one neuron to another but rather block the formation of an action potential in the downstream cell when the upstream cell is active (studied by John Eccles between the years 1930 and 1950); and the existence of the vertical columns of the cortex and the specific responses of the visual cortex (David Hubel and Torsten Wiesel, 1962).

At the same time that techniques of labeling nerve cells were being perfected, anatomical studies were producing details of neuronal connections, especially in the visual system, the motor cortex, the cerebellum (more specifically charged, it seems, with the control of motion), and the hippocampus (which certainly plays a role in memory). Electrophysiological methods of investigation allowed information to be obtained on the propagation of nerve impulses, their transmission toward specific targets, and their reciprocal reinforcement or inhibition by the impulses transmitted by nearby neurons. Not only have individual neuronal responses been explored during the past few years, but an attempt is being made to compare the response times and rates of a set of neurons having homologous functions, notably, to study their properties of synchronicity. Finally, considerable progress has been made in the molecular study of the ionic channels in neurons' cellular membranes. From this time forward we will have to think of the neuron as being the seat of highly diverse chemical and electrical activities, with time constants stretching from a microsecond to a second and even more.

■ Rhythmic Activities in Neurons

Nerve cells taken individually are in themselves typical examples of open thermodynamic systems, the ions and molecules[1] passing through their membranes and synapses constituting exchanges of energy and matter with the environment. These open thermodynamic systems are the site of temporally variable phenomena, typical of the thermodynamics of irreversible processes. In particular, the nerve cell propagates action potentials through the play of the ionic channels appropriately placed on its axon's membrane, similar to a sudden jolt running the length of a stretched rubber band (fig. 17, 18, and 19). On the other hand, the membrane potential often oscillates in a regular fashion even without the propagation of action potentials. This periodic variation, which is linked to the activity of ionic channels other

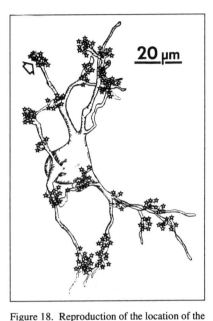

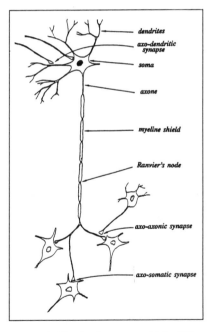

Figure 17. Sketch of a neuron with its different appendages. From *Courrier du CNRS,* nos. 55–56 (1984).

Figure 18. Reproduction of the location of the synaptic contacts between two neurons, from a study by electron microscope. Here one neuron located in the lateral geniculate body receives the axonal arborescences of a neuron whose body is located in the visual cortex. Note the disposition of the synaptic contacts into separated clusters, evoking the possibility of a specific accord between the neuronal response and the distribution in time of the impulses transmitted by the axonic fibers. Reprinted from J. E. Hamos, S. C. Van Horn, D. Raczkoski, and S. M. Sherman, "Synaptic Circuits Involving an Individual Retinogeniculate Axon in the Cat," *Journal of Comparative Neurology* 259 (1987): 165. Used by permission of James E. Hamos and John Wiley & Sons, Inc., publisher of the *Journal of Comparative Neurology.*

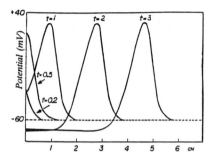

Figure 19. Propagation of an action potential in a nerve fiber. This drawing results from a computer model based on Hodgkin and Huxley's theory. The excitation of the fiber is obtained by applying an electrical current in the initial segment ($x = 0$) during 0.2 ms. Once the excitation threshold has been crossed, the action potential forms spontaneously (its state is represented at time $t = 0.5$ ms) and then begins to move in the direction of increasing x. At $t = 1$ ms its maximum has already moved by about 1 cm. At $t = 3$ ms, it has crossed 5 cm since the initial instant.

than those put into play by the action potential, demonstrates a new example of the cyclic behaviors known in open and nonlinear thermodynamic systems. Additionally, the opening or closing of many of the ionic channels in neuronal membranes, notably at the level of the axon, depends on the electric potential difference between the two faces of the membrane. The ionic current created by their opening modifies this potential, conferring on the neuron the nonlinear behavior indispensable to the appearance of dissipative phenomena.

■ The Neuron: A Logical Machine

The brain is indeed an organ of computation, which suggests the often-invoked analogy with computers, an analogy based precisely on the neurons' nonlinearity. This property allows the neurons in the logical machine that is the brain to play a role comparable to that of transistors or other electronic components that use the nonlinear properties of semiconductors.

One of the neuron's principal functions as logical machine is to make a decision by comparing two pieces of information. Reducing the possible conditions of each piece of information to a single alternative, a yes or no, means that the brain's decision must be a binary one. Possible logical binary decisions belong generally to one or the other of the following classes: "and," "or," "exclusive or," and "exclusion." For coding purposes, each "yes" will be symbolically represented by the numeral 1 and each "no" by the numeral 0, at the level both of the incoming information and of the decision to be taken. We can now make a table of the values for the principal logical functions (fig. 20). Suppose, for example, that the first piece of information corresponds to the sound of an alarm, and the second corresponds to the sight of the hands on a watch. If the brain is programmed to turn on the television when the alarm sounds and the watch reads eight o'clock, the decision will be commanded by the function "and." If only one

INF. 1	0	0	1	1
INF. 2	0	1	0	1
AND	*O*	*O*	*O*	*I*
OR	*O*	*I*	*I*	*I*
EXCLUSIVE OR	*O*	*I*	*I*	*O*
EXCLUSION	*I*	*O*	*O*	*O*

Figure 20. Values of the logical functions "and," "or," "or exclusive," and "exclusion" on two bits of incoming code. Note, for example, that if the incoming information is 0 and 1 respectively, the function "and" has the value 0, while the function "or" has the value 1, which explains the reason for the names of these functions.

of these pieces of information is enough to spur this action, the decision will be commanded by the function "or."

A neuron may receive a very great deal of information simultaneously. Some of them have more than ten thousand synapses linking them to several thousand cells upstream. The membrane potential that results from all these entries at the axon's initial segment determines whether the threshold for the creation of an action potential will be crossed. If it is, the action potential is born and propagates along the axon, reaching along the way the various synaptic "boutons" (buttons) of the axonic terminals.

In this way, the neuron compares incoming information to produce an outgoing binary signal, namely, the decision whether or not to produce an action potential. On the theoretical scale, the idea that machines constructed by humans could execute logical operations goes back to an article the English mathematician Alan Turing published in 1937.[2] The idea was revived and refined in 1943 by Warren McCulloch and Walter Pitts.[3]

Shortly after the end of the war, the first cybernetic engineers, John von Neumann especially, conceived and then began to build the first computers, which were charged with executing the calculations necessary for the development of H-bombs for the American army. We also owe to von Neumann the first analyses comparing the brain to computers, which were then called "artificial brains."

In fact, we must remark on the fact that computers are also, and even essentially, logical machines. Their basic circuits perform operations such as "and" and "or." In figure 20, note the very great resemblance between the function "or exclusive" and arithmetical addition when adding numbers written as binary numbers, that is, as a succession of zeros and ones.[4] The only difference between addition and the "or exclusive" is the process, when adding sums, of carrying over when necessary and adding that extra digit into the memory representing the power of two to a higher order. It is precisely by using this analogy that the central calculating units of computers work.

■ Natural Memory and Computer Memory

The capacity for memory offers another point of comparison between the brain and the computer. Computer memories are "addressable." They are ensembles of two-state systems (for which I will continue to use the symbols 0 and 1), "flip-flop" transistor circuits or "magnetic ferrites," whose state the central unit can question at any time by specifying the address, a number symbolizing the location of that specific memory within the total memory. At first it was thought that the neurons, being able to take two states (excited or not), represented the elementary memories in the brain, but this would mean that an excited memory, in order to register a simple 0 or a 1, would have to remain excited as long as the information remained in the memory. Today it is known that there exist plastic regions in the neurons where modifications can change the characteristics of the neuron's response given appropriate training. This is the case with synapses and also, it seems, with dendritic spines, narrow bottlenecks that sometimes link synapses to the main trunk of a dendrite.

It has been observed that the "efficiency" of the synapses between two neurons, that is, the probability that a nerve impulse would be successfully transmitted through the synaptic space between two neurons, was notably increased in some cases when the synaptic junction was repeatedly solicited. In addition, the morphological and electrophysiological characteristics of the dendritic spines sometimes changed as a function of the frequency of use of the synapses to which they gave access for the transmission of neuronal information. These changes in the synaptic regions facilitate, in a specific way, the passage of the nerve influx from one neuron to the other, but only when the junctions in question have been used successfully a certain number of times. This facilitation may last only some tens of milliseconds and disappear, or on the contrary, it may last for a long time. Thus, it is conceivable that, thanks to synaptic plasticity, preferential neuronal circuits could be imprinted in the brain after a form of "training." It is through repetition of the same "messages" traversing the same circuits, and thus using the same synaptic contacts, that these messages imprint themselves in the memory, to the extent that these synaptic contacts have their efficiency increased in a concomitant and durable fashion.

The elements of memory in the brain would, then, be found at the level of the synapses and not at that of the neurons themselves. This makes the total number of elements of memory available in the brain very great, at least 10^{13} or 10^{14}. In comparison, a personal computer of the IBM PC type generally has about 4×10^6 elementary memories. A human brain thus has a potential memory capacity equivalent to one hundred thousand IBM PCs! It seems probable, however, that only a part of the synapses present in the brain are usable for storing information, but it is not yet possible to count that proportion.

Considerable differences, however, do exist between the brain and the computer from the point of view of memory. On the one hand, the brain possesses "associative" memories instead of "addressable" ones. Rather than referring to a number to find the register in which a memory is stored, the brain can retrieve information only if it is provided with a key: a symbol or an analogy. The "central processing unit" of the brain, if it exists, does not access the contents of a specific memory by using an address, and this is one way in which it differs from a computer. Instead, in searching all available memories, it uses a kind of key, such as the idea of "grandmother," and receives in response all that this key has been able to unlock as evocations—for example, the image of the waves of her white hair or of some dessert served at her house.

In addition, memory does not appear to be strictly localized. Experiments conducted by the American Karl Lashley and going back to the 1930s have shown that a large part of a rat's cerebral cortex can be taken out without the rat losing the ability to orient itself in a labyrinth to which it has previously been habituated. If memory were localized, one would expect the rat to lose this knowledge suddenly and totally after the removal of the relevant part of the cortex. Instead, the rat's familiarity with the labyrinth is progressively diminished or destroyed as the excision of the cortical surface becomes larger. This result promoted the development of "holographic" models of the brain a few years ago.

These models take the view opposite to that of the localized-memory models, and it was suggested that memory is distributed throughout or "folded into" the cortex, like an image "folded into" the subtle correlations between the grains of silver on the surface of a hologram. But extreme positions rarely correspond to reality. Some functions of memory (learning, reliability, and resistance to minor breakdowns of the system) seem to implicate the nonlocal properties of the neuronal circuits. Other functions, on the contrary (the precision of memories, their strength, or the preservation of information over long periods of time), seem to be better explained by localized elements of memory. It is also possible that the various pieces of information relative to a single subject in the associative memory are not only dispersed in the brain but also replicated in several places, even if each element of information can be localized in one or another complex of synapses.[5]

Whatever the case, it remains difficult to determine whether memory is localized or distributed, almost certainly because we are asking the wrong questions and because we do not yet understand the concepts that are best adapted to describing neuronal memory. Progress in these domains depends not only on an accumulation of experimental findings but also on the proposition of new concepts making it possible to assemble and describe those findings.

■ The Limitations of Artificial Intelligence

The once-cherished hope that "artificial intelligence" would allow us to understand better the mechanisms of memory and human intelligence has been dashed. The analogy between the computer and the brain is, in fact, deceiving. The greatest computers cannot rival the human brain in certain synthetic chores it accomplishes in some tenths of a second. Take the example of the grandmother, who is recognizable without difficulty over a very large range of viewing angles and distances. At first sight, the superiority of biological brains over artificial computers in the tasks of "form recognition" is even more astonishing because computers are so incomparably superior in the speed of their calculation. The neuron takes a few milliseconds to complete an elementary operation and transmit the result to the next cell: the biological signal or action potential lasts a millisecond or more and propagates with a speed varying between ten centimeters per second, in the dendritic terminals of the cortical cells, for example, and some tens of meters per second, in the case of the large nerve fibers with an insulating sheath of myelin. In contrast, the logical register of a computer performs an elementary operation in a fraction of a millionth of a second, and information circulates in the computer at the speed of light, about three hundred thousand kilometers per second. In the few tenths of a second necessary for recognition of a face, the visual information certainly does not go through more than about ten steps of "computation" or cross more than about ten synaptic doors.[6] In this same period of time a modern computer can perform several hundreds of thousands of elementary operations. To arrive at the same result as the brain, however, to identify, for example, the grandmother's face from any angle whatsoever, it would take a computer several hours of calculation and information sorting. Nonetheless, the machine's advantages become apparent when the computer is measured against the human brain in the tasks at which it excels, such as scientific computations. Then it shows itself to be incomparably more rapid than its biological homologue.

The fascination exercised by the computer, which resulted in the development of artificial intelligence over the years 1960 to 1980, certainly helped to retard the progress of theoretical neuroscience. Rather than first seeking to understand how the brain truly functions in order to imitate it, researchers used the available computers to simulate intellectual functions. Intoxicated by the first encouraging results, the natural tendency was to go further and explain the mechanisms put into play in mental faculties. This was a strange undertaking. It would be comparable to the attitude of a researcher in 1980 asking aircraft manufacturers to explain how and why birds fly and pressing them to specify, for example, the avian equivalent of jet engines. Having come up against unexpected obstacles, researchers are now turning toward new theoretical analyses touching on the functioning of the nervous system and are looking more closely at the findings of neurophysiology. The new strategy suggests that it is necessary first to integrate these findings with the models of central nervous system function and only then to develop these models and construct machines specialized in "neuromimetics."

■ Parallel Processing of Information

At present the whole of the scientific community does not recognize any one reliable classification of physiological findings relative to central nervous system function. The most important and significant from the point of view of the brain's temporal functions are presented below.

First, the brain presents a highly parallel architecture in which "cooperation" must play an essential role. This means that, when a piece of information moves from one point in the brain to another, there is not generally one direct nervous fiber charged with specifically ensuring this link. Most often, a given neuron contacts numerous downstream neurons. Conversely, this neuron can receive and integrate the responses of hundreds, even thousands, of upstream neurons. In addition, Lashley's studies of neurophysiology showed that memory is not, contrary to the case of computer memory, located at a very precise address. The global behavior of a group of interconnected neurons used in parallel is as important, if not more so, than the particular lines of neurons the information follows. It can be said that the "network" aspect is more important than the "cabling" aspect.

Forced resonances must surely establish themselves in such networks, synchronizing the activity of vast ensembles of neurons. The simultaneous discharges of these neurons in repeated bursts can give birth to oscillations of electrical potential of sufficient amplitude to be detected even outside the brain. This is indeed the case with the waves recorded in electroencephalograms using electrodes resting on the surface of the scalp. The alpha waves, the most regular, are observed in conscious but relaxed subjects whose attention has not been solicited. They are probably born in the reticular formation at the base of the brain, and it does not appear possible to create them except by synchronizing the nervous discharges of millions of neurons. Their period of about a tenth of a second could correspond to the time needed to analyze the impressions received or to recapitulate at regular intervals the internal state of the brain. In sum, the cerebral waves could orchestrate different stages of intellectual activity. It has been observed, in fact, that the sensitivity of temporal perception in monkeys and in humans depends on the phase during which the stimulation is applied. Subjects discriminate two flashes of light separated by a very short time interval better when the stimulation comes during the increasing phase of the alpha waves.[7]

Other mechanisms of cooperation come into play to select a particular network of neurons under the influence of a given stimulation. The use of such networks allowing information to take multiple routes explains why the brain resists minor breakdowns of certain of its "components" so well. Some researchers estimate that an adult human loses about a thousand neurons each day, but his or her cognitive functions do not appear to be spectacularly affected by such a loss!

The computer simulation of neuronal networks was, it seems, begun by von Neumann, who wanted to verify the reliability of the networks and "parallel" structures. In these structures the transformation of input into output is entrusted to numerous similar elements linked in parallel, not to a single element of calculation. Today the "connectionist" models derived from this study have obtained unhoped for results in many other domains. They shed new light on the problems of training and pattern recognition, so much so that new machines adapted to these tasks, called "neuro-computers," are beginning to appear on the market.[8]

Research on neuronal cooperation and the selective activation of networks within the framework of connectionism is in full-scale development and is already showing spectacular results. For example, at the Massachusetts Institute of Technology, Terrence Sejnovski was able to teach NET-TALK, a network of artificial "neurons" equipped with a loudspeaker, how to read a text in English and to pronounce it correctly. This result was obtained after several training sessions, without any specific program being given in advance to the machine. Note that, as was the case with the first studies on the computer, this technological success stands in contrast to the relative rarity of solid indications regarding the biological brain's use of corresponding principles. In addition, most of the models used today give a static image of the brain. They suggest how memory can be distributed in the brain and restored in particular conditions of stimulation, but they do this without clarifying its dynamic functioning. The situation is comparable to the beginnings of wave mechanics, which allowed spectra of the energy of stable states of systems to be obtained but did not describe their evolution over time. In any case, new theories have recently appeared that emphasize dynamic cooperation of neurons and implicate a synaptic plasticity with a short time constant. Some of them seem to lead to interesting results in the field of form recognition under continuous change (as in the problem of recognizing the grandmother's face from any angle).[9]

■ Preparing for Action

The brain must constantly compare the sensory information furnished by sight, hearing, and touch to transform it into motor commands. Because the senses and the motor organs use different natural coordinates, comparisons between data from different senses, as well as their transformation into decisions affecting motor acts, involve the transformation of coordinates, in the geometrical sense of the term. We are beginning to understand why the networks of neurons are well-adapted tools for these transformations[10] and how these mechanisms involve the existence in the brain of vaguely homeomorphic maps of the sensory findings. In other words, they produce a projection that is more or less similar in topography to the external stimulation: retinotopic maps of the visual cortex, the lateral geniculate bodies, and the superior colliculus; auditory maps; and sensorimotor maps of the cortex, all along the central sulcus, and so on.

For example, the most natural system of coordinates for vision is a system whose axes are linked, on the one hand, to the spherical geometry of the retina and, on the other hand, to the state of tension of the oculomotor muscles. The most natural system of coordinates for the arm or leg is polar, inasmuch as the position of the extremities of these limbs depends on the angles of rotation—or of the tension of the flexor muscles and their antagonists—of the various joints. The brain must thus execute a transformation of coordinates between the visual perception of an object and the motor command for the arm to reach it.

In analytic geometry a transformation of coordinates is accomplished by multiplying the vector representing the original coordinates by a matrix whose elements specify the type of transformation to be made. The resulting vector describes the new coordinates. Neurons are well suited to this type of calculation. In fact, in matrix multiplication, the components of the new vector $\mathbf{N}_i$ are obtained by a weighted sum of the components of the old vector $\mathbf{A}_j$ with the weights $\mathbf{B}_{ij}$ given by the transformation matrix ($\mathbf{N}_i = \Sigma_j \mathbf{B}_{ij} \mathbf{A}_j$). In a neuronal ensemble the output activity of the neuron $\mathbf{N}_i$ is also a weighted sum of the various activities of the neurons of entry ($\mathbf{A}_j$), the weights being fixed by the efficiency $\mathbf{B}_{ij}$ of the synapses linking neuron i to neuron j. The coefficient $\mathbf{B}_{ij}$ can be, of course, either positive or negative. The negative coefficients might correspond to inhibitory synapses in the brain.

■ Coding Information

In a given nerve fiber traversed by a train of "spikes" (electrical impulses), information must be coded temporally. The nerve impulses transported by a single axon are all of the same size and form. As a consequence, it is not through modulating this form that the neuron can transport information. The neuron must play on the intervals, the silences, separating the successive impulses. From the local point of view the brain must use a system of coding closer to a modulation of frequency than of amplitude.

Until recently it was thought that information locally transported consisted simply of the nerve cells' average frequency of discharge and that the temporal arrangement of these discharges was purely stochastic. In other words, it was thought that only the number of impulses per second, and not the distribution in time of these impulses, had a meaning in a neuron's response to a given stimulation. But a local code founded exclusively on the average frequency of the discharges would not be flexible enough to account for the brain's capacity and reliability.[11] Furthermore, research in progress seems to show that the amount of information coded under the form of temporal correlations in a sequence of nerve impulses—beyond the simple mean frequency of discharge—increases regularly in the visual system when the information passes from the optical nerve to the lateral geniculate body, to the visual cortex, and finally, to the temporal cortex, which seems to house the principal circuits involved in the recognition of faces.[12]

Beyond being encoded in the average frequency of discharge, information could be transported and coded in the temporal structure of trains of spikes, in the silences of variable length that separate the different nerve impulses, a bit like morse code. This code, which was once used for the electrical transmission of telegrams, employs the relative length of the signals, dots and dashes, to distinguish the letters of the ordinary alphabet. According to this hypothesis, then, one could detect information transmitted from place to place in the brain along a given route by very closely studying the temporal correlations between the various impulses constituting a neural "response."

At Los Angeles, in 1969, B. L. Strehler suggested that groups of specialized synapses could use spatial summation to detect and sort from among a train of spikes those that are specifically directed at the cell to which those synapses belong. The cell would not fire unless the impulses transiting the same afferent fiber, or fibers belonging to the same axonic arborization, fell simultaneously or almost simultaneously on each of the synapses in question. It is supposed that an additional mechanism acts in conjunction with this one—the temporary facilitation of the synapses.

These would eventually allow the filter, which would comprise a defined group of synapses, to retain, memorize, and transmit only the temporal impulse patterns conforming to the original pattern. Specialized information could thus be directed toward precise targets.

According to the proposed model, information could be coded in a multiplex, redundant fashion by short motifs, such as triplets of impulses, that are repeated in the same train and recognized by specialized groups of synapses. The model would also explain how several pieces of information could be memorized in a single group of synapses and how a precise piece of information could be selectively extracted from that group. Very recent studies seem to show that such precisely repeated patterns do, in fact, exist in the sequences of impulses registered in the brain, at least in the region of the visual cortex, the only one to have been studied in this respect until now.[13] But the precise function of these repeated patterns is still not known.

■ Coordinating and Directing the Organism

The clock is the essential component of a computer, at the very heart of the central processing unit. It commands the step-by-step execution of the program, the loading of the registers, their comparison using a logical function, and the storage of the results into memory. On this point at least, the most recent models of the brain follow the artificial archetype. In a connectionist model, for example, the state of the neuronal network is examined at discrete intervals, step by step. At time 0, the external stimulations are represented by the imposed activities of certain nodes in the network representing the neurons. The subsequent state of the network resulting from interneuronal communications is then examined. The synaptic weights (that is, the efficiency of transmission from one node of the network to another) are adjusted, and the cycle begins anew. Models in which the information is coded in the form of temporal correlations between successive impulses require a time of minimum integration, so that all the correlations can be "decoded" and the "message" understood or stored in the implicated synaptic ensembles. In this case, in fact, the message consists not of an instantaneous or quasi-instantaneous event, such as a single impulse, but of groups of impulses distributed in time and possibly repeated.

The brain contains many clocks, and we have learned to identify some of them. A single master clock probably does not exist. Given the great flexibility provided by the parallel functioning of the brain, it does not appear necessary that all logical operations be directed by one single clock. Rather, they could be controlled by several decentralized clocks. In other respects, the brain waves seem to give rhythm to the brain's functions of integration and successive "computation," just as a computer's central clock imposes its rhythm on the different stages of a calculation and on the succession of logical steps effected. In 1988 it was discovered that the neurons of the visual cortex that respond to stimulations having a logical relationship to each other (for example, these neurons are sensitive to luminous segments belonging to the same contour) have a tendency to discharge themselves in a synchronous fashion, in "coherent oscillations."[14] But the definitive proof of the function of integration and successive computation of these oscillations is still missing.

■ Thinking

Finally, the brain is the physical support of conscious thought. It must, thus, also furnish an internal representation of the world and the self, a representation involving decisions on the sense data that are worth keeping and being used to construct the "ego," the unity of the self and the coherence of the external world. This unity of the self and coherence of the external world are manifestly not immediate data. They depend, on the contrary, on a delicate and complex cerebral activity.

Although we are far from knowing the mechanisms of this construction of the self and from classifying all its aspects, it seems certain that the temporal functions of the brain play a very great role. We have just seen, in fact, that the brain does not instantaneously carry out the activities of perception, memorization, and recall, which leads us to conceive of its activity as being divided into distinct "moments." In the function of conscious perception, however, these diverse moments of immediate perception must be reunited and blended. In this way the fluidity of time's passage can be reconstituted, since conscious, waking thought undeniably includes a particular function charged with appreciating at each instant the "time that passes." Bergson calls this time "pure duration, which is the form that the succession of our conscious states assumes when our 'ego' lets itself live."[15]

■ Neurological Time and Psychological Time

Research thus leads us to and invites us to cross the still-unbridged chasm that separates experimental neurosciences from experimental psychology.

Paul Fraisse, in his study *Psychologie du temps* (The psychology of time), insists particularly on *two* critical durations of perception in humans. Beyond the key frequency of 10 Hz, successive impressions blend into one another. Two images presented at an interval of 100 milliseconds form a composite image. In films this fusion is used to restore the illusion of motion. This is also the maximum rate at which one can strike a piano key, thus also the maximum at which successive and distinct motor orders can be given. Paul Fraisse himself points out the coincidence between this critical duration and the period of alpha electroencephalographic rhythms.

The second critical period, of about 600 ms, corresponds to the spontaneous interval between a stimulation and an elaborated perceptual response, such as naming an object. Since this is also the interval of duration that can be perceived with the most accuracy, it thus constitutes, in some way, the natural "tempo" itself, used as such in musical scores. Whatever the case, and even if the estimated durations may be appreciated in diverse ways and vary from one individual to another, their existence and the segmentation of psychological time that they imply do not seem to be in doubt.

The idea of a segmentation or, in the terms used by physicists, a quantification of psychological time is not new. It was already to be found in the writings of William James, who, it seems, was the first Westerner to discuss this discreteness of the mental mechanisms of perception, whose "moments" must follow each other as fixed images do in a film. "Our acts of recognition or of apperception of that which *is* must be discrete," he wrote.[16] Our feeling of continuity in such a succession, which we call duration, must thus be an elaborated cognitive function, it being understood that "a succession of feelings, in and of itself, is not a feeling of succession."[17]

■ The Acceleration of Lived Time

We experience ourselves living, and we sense the time that passes. But do we not feel that this time, lived time, does not flow uniformly? Sometimes it drags out, as a measure of our boredom or impatience to provoke an event or to go to meet it. At other times it passes so quickly that we have difficulty following it. Someone once joked that death comes when we can no longer run quickly enough to keep up with time, when it decides to pass us by once and for all. We are almost unanimous in judging that psychologically, if not even biologically, time passes faster and faster with age. One month at school seemed to us an eternity. One month of work at a mature age seems frightfully short.

The physiological foundations of such an impression are very poorly understood. Insofar as it is difficult to quantify subjective impressions and to construct the experimental protocols that would allow their comparison in a scientific or reproducible manner, psychologists have not yet reached conclusions on this subject. James thought that *"the same space of time seems shorter as we grow older*—that is, the days, the months, and the years do so; whether the hours do so is doubtful, and the minutes and seconds to all appearance remain about the same." James probably meant that the impressions of older and younger persons do not differ substantially when the psychological test of estimating an interval of passing time is applied to brief enough intervals. But older persons can, in taking these sorts of tests, correct their immediate and intuitive sense of the "speed" of time's passage. In fact, it must be recognized that despite the exacting research of Piaget, Fraisse, and contemporary psychologists, little is known about the psychological perception of duration and the relationship between psychological time and biological clocks.

Could not the impression that time runs faster be linked to the progressive slowing down of a biological clock? The eventual deceleration of alpha rhythms during aging could a priori explain the psychological impression of an acceleration of time (the brain would use the period between two peaks of alpha wave as a time standard, and the same lapse of physical time would be measured by a smaller number of periods in older persons than in younger ones). A slight diminution of the frequency of alpha rhythms has indeed been observed in older persons, but this lessening, on the order of only a few percentage points, is much too small to account, in this way, for the psychological impression of an acceleration of time.

■ Psychological and Present Time

The importance accorded here to a segmentation of psychological time is not fortuitous. It seems necessary in order to understand and give meaning to a phrase that is fundamental to our lives, "the present." The time of which I have been speaking until now, which is observed in action in the cosmos, in physical nature, and the biological world, is manifestly endowed with an arrow. But, for all that, it is not segmented into the three familiar categories of *past, present,* and *future.* Reworking a famous phrase of Einstein's, the world is not in all respects truly deployed in all its temporal dimension; it is actually "in the process of becoming." Still, the separation between past, present, and future could be no more than "illusion."[18] Physical and biological time are certainly oriented, but they do not include this privileged slice we call the present. Being essentially continuous, time can be divided—as Aristotle himself noted—between the before and the after, but the "instant" that separates them can no more be a piece of time than a point can be confused with a piece of a line.

Topologically, the markers of a partition, such as instants and points, cannot be of the same nature as that which they divide. Briefly, if the present is for us a mobile bit of time, it cannot be other than a sort of "thick" present," unlike the mathematical instant, which does not have a temporal extension. Such a distinction between a thick present and an instant without thickness had already been noted by Edmund Husserl and Charles Peirce.[19]

Thus, the present corresponds to an interval of time in some way, including at least one unit of psychological time, if not more. In fact, it is the piece of time that is always mobile, during which we become conscious of sensations that reach us and we prepare the actions that we are about to undertake. From the psychological point of view, therefore, the "present," *stricto sensu*, is perhaps not as important as the immediate future. It is the immediate future that, through the actions we prepare in the present instant, gives us an acute sense of the present, and not the reverse. Bergson, even before Heidegger, already held this opinion when he wrote that we judge a thing's degree of reality by its degree of utility: "our representation of matter is the measure of our possible action upon bodies."[20]

■ The Future Horizon

Biologically, the extraordinary development of the human frontal lobe over the past million years seems to correspond to the evolution of our faculty for anticipating the future, for looking always farther ahead. The specificity of human psychological time thus rests more, it seems, on our capacity to predict the future than on our ability to remember the past. Memory is necessary to anticipate the future, but it is not enough. Animals have an elevated faculty for memorization, but they have only mediocre aptitude in exercises of long-term anticipation. One of the essential characteristics of human cognitive functions thus concerns the temporal horizon that, no doubt in relation to the development of the frontal lobe, has considerably expanded toward the future. Here, however, we run into the problem of the difference between the appearance of the organ and the appearance of the function. Although the development of the cerebral cortex goes back to about a million years, the capacity for autoprojection toward the distant future seems to date only from about ten thousand years. The archaeological traces left by past civilizations suggest in fact that in the prehistoric cultural universe, before this period, the metaphysical anguish over a distant future did not exist. What happened in the meantime? What was the psychology of time in our ancestors *Homo erectus* or *Homo habilis?* It is possible to suppose that a slow *reorganization,* a progressive programming of the brain in its intellectual function, succeeded the explosion in cerebral *capacity.*

■ ■ ■

Now we are able to discern the major outlines of the role time plays in neuroscience. We already knew about the mechanisms of learning and transmission of nerve impulses, and it is certain that these are used in the cognitive functions of the brain. If they are not yet entirely understood, the mechanisms of memory are at least beginning to be unveiled. It is thought that they involve changes of the physico-chemical state of the neurons, particularly in their synaptic regions. These changes could mean that during recall, a configuration of neuronal activity identical or close to that which was linked to the initial event might be reconstituted. Finally, although it is still unknown how these signals can be effectively used for logical reasoning and voluntary behavior, the current models that attempt to explain them increasingly emphasize the preeminent role time plays in these functions, through the synchronicity of the signals or the repetition of the temporal patterns they constitute.

If the consciousness of self reaches a qualitatively higher level in humans than in animals, it must indeed be due to this temporal faculty of the brain and its orientation toward the future. Martin Heidegger underlined the organic links between the "I" and anticipation. "The essential phenomenon of time is the future," he wrote. "In order to understand this without trying to phrase it as an interesting paradox, the perpetual *'Dasein'* (being-there) must maintain itself in its anticipation."[21] Here surely, we must recognize the intuition of a profound truth, touching precisely on mankind in our relationship with time. Our ego, our I, is founded on the conviction that, in the next instant, in a minute, in a day, or in a year, we will be the same and will always pursue the same satisfactions, the same ends, and will nourish the same fundamental desires. Physiologically, we are perhaps essentially "neuronal," but psychologically, we are, above all, "temporal."

Epilogue

The scientific adventure, which was begun essentially less than four hundred years ago, has already allowed us to turn our backs on certain myths (the eternal cycles, unequal hours, and, more recently, absolute and universal time). We have been able to explore successively the two principal and complementary facets of the notion of time by creating the tools for that exploration, the concepts of causality and of entropy.[1]

But questions about time, causality, and entropy have in their turn outgrown the framework of scientific constructions. They have been at the heart of questions attacked by the most astute philosophers, armed with the only knowledge available at their time. Among these, Kant, Hegel, and Bergson are incontestably preeminent. It might be helpful now, in closing this discourse on time (and not coming to any definitive conclusions that are beyond the range of the scientific method, particularly on a subject such as this one), to compare the teachings of contemporary science with the intuitions of these three thinkers.

■ Taking Another Look at Causality

The theory of relativity confirmed that time never runs in reverse. This is true for mechanical clocks, whether natural or artificial, as well as for thermodynamic clocks that use the radioactive properties of atoms or biological clocks whose internal rhythms dissipate entropy. Physical causality accords with the definition of time as a degree of freedom. Objects using it retain their identity while displacing themselves in a direction called "future," *which always stays the same.*

But the theory of relativity crowns a much more precise vision of time than the property of nonreturn might suggest. It makes of causality a universal paradigm of order subjected to the invariance of the speed of light, which takes the value c in all possible systems of reference. The universality of this paradigm led physicists to develop a causal interpretation of time.

Today, however, it is no longer possible to reduce time to its causal interpretation. The laws of quantum mechanics, Heisenberg's uncertainty principle, and the principle of superposition of quantum states and the phenomena of inseparability that come from it have shown the difference between reality and this rule of thought. To return all the power of a universal paradigm to causality, it will be necessary to find a more subtle definition for it than the Marquis de Laplace wished to give it. Since determinism does not have this universality, which would have allowed it to be identified with time, the future of the world is perhaps not entirely written in its present state.[2]

This perspective brings us closer to Kant, who professed that space and time are not properties of nature in itself and that causality does not hold sway over it.

■ Kant and the Restoration of Liberty

For Kant, space and time are forms of intuition—that is, innate ideas that allow us to order the world—but they are not entirely of nature. They reflect, in part, the structures of our minds. In addition, our perceptions give us only an already interpreted version of the world through precisely these categories of space and of time. It is only in this interpreted version of the world, that of "phenomena," that causality applies.

Science does not say exactly this. Reality in itself is certainly subject to a law close to that of causality, even if it is not the causality proposed by the Einsteinian theory of relativity. Faced with the absolute and "divine" time of Newton, Kant rightly insisted on the cognitive, psychological, and epistemological aspect of this notion. Our time is not the raw time of nature. It is located midway between things and us, closer even to us than to things. But ever since we learned to dissociate time and causality and have known causality's limits of application, it does not appear necessary to return to the Kantian artifice to open a breach in nature through which liberty could win entry. Liberty is, perhaps, no more than a particular property of ontological time, a sort of profound conformity with the nature of chance that is not as blind as might have been thought. Spinoza and Leibniz already saw in liberty an acquiescence to the profound fluxes that traverse nature. Scientific epistemology could very well agree with them on this point tomorrow.

■ Hegel, or the Dialectical Duration

Hegel made becoming the "form of being of nature," and in this the recent orientation of science is leading us closer to him. At the same time, however, Hegel denied reality by enclosing it in its dialectic as in a stranglehold. Far from succeeding itself, reality is born at each moment from nothingness and returns to nothingness.

> The dimensions of time, *present, future,* and *past,* are the *becoming* of externality as such, and the resolution of it into the differences of being as passing over into nothing, and of nothing as passing over into being. The immediate vanishing of these differences into *singularity* is the present as *Now* which, as singularity, is *exclusive* of the other moments, and at the same time completely *continuous* in them, and is only this vanishing of its being into nothing and of nothing into its being.[3]

The absolute idealism of the Hegelian approach does not in any way allow nature to be made the object of a scientific inquiry in the frame of realism or, in particular, objective properties such as causality to be attributed to time or to becoming. Once again, traditional causality appears clearly inadequate to represent an objective property of nature. Nevertheless, a certain form of causality must be accepted in order to investigate the true nature of becoming with the instruments of science or philosophy. Without it, the present could not teach us about the future and science would be powerless. Even worse, the present could tell us nothing about the present, since, as Hervé Barreau recently remarked, "becoming . . . is the change which the whole of reality undergoes and exercises in the present."[4] Psychology agrees with this conclusion, which sees in our present, in accordance with the Bergsonian analysis, the entirety of possible actions our bodies could perform on our images of the real.

■ Toward an Objective Entropy

Progress made in the comprehension and refinement of the second principle of thermodynamics[5] and its role in nature helps us to know more about the nature of becoming. With the concept of entropy, we try to affirm the absolute character of becoming, the reality of change in the world (whatever is, in the final analysis, the ultimate substance of the world). Not long ago most physicists ranged themselves resolutely in the school of "realism," which professes that physical theory has to do with reality in and of itself, independent of any observer. In view of the identification between time and causality, these same physicists taught that mechanistic determinism and the relativistic rules of the propagation of causes applied to reality itself. On the other hand, they did admit that entropy lacks a purely objective significance, for the simple reason that this concept describes the evolution of systems toward states that are more and more "probable," the notion of probability being undeniably subjective. It is true that entropy maintains a strict relationship with the notions of order and information, notions that almost everyone agrees cannot be attached to the real itself, independently of any subjective judgment (because order is judged in relationship to a project, to a particular intention) or observer. But are order and disorder always relative to a project? It is possible to conceive of absolute disorder: a disorder that is not affiliated with any project and for which no Rosetta stone exists. Perhaps the substance of the world arises from a cosmic chaos whose principal and only property would be of allowing no order to be defined by any observer whatever and would therefore, in that sense, be an objective chaos. Thus a subjectivism of causality (at least as it is defined in the handbooks of relativity) and a realism of entropy, or of something resembling it, might replace the causal realism and the subjectivism of entropy of the recent past.

It is in the play between time and chance that new progress in the exploration and comprehension of time should be expected. To raise a new corner of the veil, we still need a clear view of the difference between the concrete chance that is at work in the world and the mathematical chance that is abstract and evidently incapable of leading the universe, objects, and beings toward their destinies. Time resides perhaps in this difference between mathematical and concrete chance. It may be a minimal difference in appearance, but, by working on the 10^{80} particles that populate the universe, reigning as master of the void itself and especially of the original void that began everything, it can forge becoming. We may, then, hope to understand better the fundamental philosophical problem of the exact relationship between determinism and innovation and give a meaning to Bergson's statement that "time is invention or it is nothing at all."[6]

■ Matter, Duration, and Life according to Bergson

Taking a position opposite to that of Kant, Bergson admitted that time exists at the heart of matter. Far from being inaccessible, he said, it is immediately accessible, not through intelligence, but through intuition. It is true that intelligence constructs concepts and that these concepts structure the real. "But the truth is that our mind is able to follow the reverse procedure. It can be installed in the mobile reality, adopt its ceaselessly changing direction, in short, grasp it intuitively," he wrote in *La Pensée et le mouvant*.[7]

In many respects, it is impossible not to be struck by the accuracy of this philosopher's view. Already, fifty years ago, he proclaimed that science martyred and denatured time by projecting it into space (which the causal interpretation of time does). On the other hand it must be recognized that Bergson, rightly insisting on becoming, did not fully appreciate the importance of the other facet of time, that which is linked to causality. Not understanding its true implications, he denatured the theory of relativity in his book *Durée et simlutanéité* (*Duration and Simultaneity*).

Bergson was certainly right to insist on the flaws of determinism. His intuition did not betray him when he proclaimed that time is invention. But he did not understand that life is not a constant effort to climb the slope of entropy. There is no dichotomy between matter and life: both obey the principles of thermodynamics. Although the first usually breaks up and degenerates to uniformity by means of these principles, the second draws from them the strength to organize itself. Invention, which Bergson says duration has the right to claim, is found thus at the very heart of entropy. To unmask it, chance, or rather universal chaos, must be further examined.

What can be thought of the dualism Bergson introduced, of the break he allowed between life and thought? This question will remain outside the field of scientific inquiry for a long time to come, even though it can be predicted without much risk that this frontier between science and religion will be pushed back. To the extent that we understand the role of time first in physical nature and then in the deployment of life, in the flourishing of individuals, and finally in thought, we will better discern the limits of the realms that the scientific method opens to our exploration. The science of time already allows us to understand the mechanisms of regulation of the living. Tomorrow it may permit us to understand memory. The only thing that will then escape us will be that which represents the supreme summit of thinking life: mental representation and the free act.

■ Between the Cosmos and Life, the True Face of Becoming

The power of time is affirmed when the ladder of complexity is climbed, toward either the infinite space of galaxies or the microscopic complexity of living beings. Our analysis must take these into account. Is it chance or a sign if the most complex being is also the only one to possess a sense of time sufficiently honed to feel abstractly the anguish of its mortal condition, even in the absence of any immediate danger? This existential anguish, which expresses the absurdity of our condition, was well described by Albert Camus. Is it not the hope for freedom from its clutches that pushes us to an impassioned search for a total comprehension of time?

Science henceforth permits us to think that becoming, whose power we can see increasing from the atom to the star, from the elementary particle to the living being, in some way constitutes the true sap of the real. Time is not only a decoration of life, a system of coordinates, or the conductor of an orchestra. It works more deeply. It is the soul itself of matter. It is thus not only an idea, not even an Idea. It is not only the product of the mental processes of our brains, or something in which, so to speak, existence would be separated from substance.

The Big Bang and the majestic flight of the galaxies in the universe, the evolution of living beings from the amoeba to humankind, and the birth of an infant and its growth toward maturity are the most astonishing manifestations of time-becoming. Evanescent at the level of the elementary particle and the individual atomic nucleus, difficult to apprehend with the tools of analytical science at the scale of the physical systems of everyday events, it does not take on its full texture except at the extremities of the scale of complexity, whether it is a question of the spatial complexity of the universe or of the structural complexity of living beings. These two extremes meet each other in the significance of our search for an understanding of time. Is it chance or an omen if we have repeatedly encountered a parallelism between the destiny of the world and that of life?

Notes

Introduction

1. H. Bergson, *Time and Free Will,* p. 100.
2. Between 1960 and 1980 R. Sperry studied several patients who had undergone surgery to sever the corpus callosum, the batch of nerve fibers responsible for transferring information between the right and left brains. The operation was aimed at reducing or eliminating the recurrence of epileptic seizures resistant to nonsurgical treatments. For a description of this work and the psychophysiological conclusions that can be drawn from it, see P. S. Churchland, *Neurophilosophy* (Cambridge: MIT Press, 1986), p. 174.
3. The relatively primitive eyes of some frogs are incapable of perceiving a motionless fly but quick to locate it in flight.

Chapter 1: The Myth of the Eternal Return and the Concept of Progress

1. J.-P. Vernant, *Myth and Thought among the Greeks* (Boston: Routledge and Kegan Paul, 1983).
2. Plato, *Plato's Cosmology: The Timaeus of Plato,* trans. Francis MacDonald Cornford (Atlantic Highlands, N.J.: Humanities Press, 1952), 37c.
3. A. Koyré, *Galileo Studies,* p. 67.
4. Aristotle, *Physics,* trans. H. G. Apostle (Bloomington: Indiana University Press, 1969), 220a.
5. Ibid., 223.
6. The distinction between the supralunar world and the infralunar world certainly was a sign of progress in the emergence of the notion of physical time, because the celestial spheres' perfect motion with which the ancient philosophers were satisfied did not prepare them in any way, either before or after Aristotle, to grasp the importance of the principle of inertia, the only correct means of returning time to a measurable physical dimension (and no longer only enumerable, as with Aristotle). In a much later reversal of thought, Newton would use precisely the principle of inertia, established as a principle through the study of concrete motion, to explain celestial movements.
7. Saint Augustine, *The City of God,* book 12, chap. 13.

8. H. Barreau, "La Construction de la notion de temps" (The construction of the notion of time) (thesis, University of Paris X, 1982).

9. J. Locke, *An Essay Concerning Human Understanding*, p. 203.

Chapter 2: The Invention of the Concept of Instantaneous Velocity

1. J. Piaget, *The Child's Conception of Time*, p. 268.

2. The history of dynamics in ancient and medieval physics is well reported in A. C. Crombie, *Augustine to Galileo*. The law proposed by Aristotle to explain the differences in observed velocity between diverse motions is the following: the heavier the body or the more force employed, the faster a given distance will be traversed. In another instance, the larger the resistance of the traversed medium, the slower the motion. In modern mathematical terms we would write $v = k(F/r)$. The velocity v is proportional to the force F and inversely proportional to the resistance of the medium r, with k being the proportional constant. Of course, Aristotle didn't write this equation, since for him, as for all the Greeks, there was no way to measure the relationship between two magnitudes not measurable by the same standards, such as "force" and "resistance." For the same reason, he did not describe velocity as a relationship between distance traveled and the time necessary to travel it.

The law proposed by Aristotle would later be modified for several reasons, in particular by his Arab critics, who denounced as absurd the idea that a moving body could attain infinite velocity. That is, however, exactly what Aristotle's law predicts when the resistance of the medium is nullified, for instance, in a vacuum. And if the resistance of the medium surpasses the force imposed on a body, the latter will remain at rest, although Aristotle's law still predicts a nonnull velocity.

For these reasons, Averroës and Avempace proposed another formula in which velocity is proportional to the difference between the propulsive force and the resistance rather than being proportional to their ratio: $v = k(F - r)$. Later, the physicists of medieval Europe, such as Oresme, Buridan, and Albert of Saxony, would rework Aristotle's formula to take the second objection into account. They proposed the law $v = K\ln(F/r)$, in an implicit form of course, since this was before the invention of logarithms. This law indicates that the body is at rest when the medium's resistance is equal to the force imparted to the body.

3. Koyré, *Galileo Studies*, p. 149.

Chapter 3: Galileo's Remarkable Error

1. Koyré, *Galileo Studies*, p. 149.

2. William R. Shea, *Galileo's Intellectual Revolution: Middle Period, 1610–1632* (Canton, Mass.: Science History Publications, 1972), p. 89.

3. P. Thuillier, "Galilée et l'expérimentation" (Galileo and experimentation), *La Recherche* 14, no. 143 (1983): 442.

4. Galileo, letter to Paolo Sarpi, Feb. 16, 1604, cited in Koyré, *Galileo Studies*, p. 67.

5. Galileo, Discourses, Third Day, cited by D. Dubarle in *Galilée: Aspects de sa vie et de son oeuvre* (Galileo: Aspects of his life and work) (Paris: P.U.F., 1968), p. 252.

CHAPTER 4: NEWTON AND THE DISCOVERY OF THE LAW OF GRAVITATION

1. Newton knew that with a sling the centrifugal force is proportional to v^2/R, where v is the velocity of the object turning at the end of a cord of length R. The velocity of a planet in a circular orbit whose radius R is $2\pi R/T$, where T is the period of the revolution that is observed. The force exerted on it must therefore be proportional to R/T^2. In addition, Kepler's third law, which was already known at this time, indicates that the square of the period is proportional to the cube of the orbit's radius. The force sought is therefore proportional to R/R^3, that is, to $1/R^2$.

2. The altitude of a body in free fall varies as $y = \frac{1}{2}gt^2$, where g is the gravitational acceleration and t is the time elapsed. In the first second of the fall, therefore, this body nears the horizontal plane of the quantity $g/2$. Consider the circular trajectory of the Moon when it is at its zenith. In the following second, it approaches the horizontal plane by the quantity $R(1 - \cos 2\pi/T) \cong R/2(2\pi/T)^2$, where R is the radius of the lunar orbit and T the period of that body's revolution around the Earth. This quantity must therefore be equal to $g'/2$, where g' is the gravitational acceleration at the altitude of the Moon. The law of universal attraction gives to the latter the value $g' = g(r/R)^2$, where r is the terrestrial radius. Combining these diverse equations, one obtains the relationship sought: $T^2 = (4\pi^2/g)R(R/r)^2$.

3. Newton could have, in fact, contented himself with requiring the laws of mechanics to relate to a privileged reference system, that in which the axes point to "fixed stars," without affirming anything about the existence and the primacy of absolute space and time. Indeed, what is an empty space—empty but evolving in time—if it is not a "thought" space? Affirming the existence and the primacy of space over matter is to suppose implicitly that there exists a thinking entity to conceive of this space. Newton's predecessors hotly debated whether space and time could exist independently of the material world, and Descartes, forty years before Newton, still professed that "that which we call time is nothing, outside of the true duration of things, but a manner of thinking" (Descartes, *Principles of Philosophy* [Boston: Reidel, 1983], chap. 1, sec. 57).

4. In contemporary physics, a particle is virtual when the energy it transports is insufficient to give it a true existence, that is, an observable existence, with its own characteristic mass. It remains therefore hidden, so to speak, in the stuff of the world, without reaching the surface of phenomena.

5. St. Augustine, *The City of God,* book. 11, p. 350

6. In the *Principia* Newton explains that absolute space and time are as the impalpable fabric with which God drapes himself, or by which he manifests himself. In a correspondence between Clarke and Leibniz, Clarke develops Newton's idea and compares the perception of ideas in our minds to the perception of things by God and describes absolute space and time as the divine "organ" of this sensation.

Chapter 5: Time and the Natural Sciences

1. I have already noted that, not having placed time at the heart of his scientific reflections, Leonardo da Vinci missed beating Galileo in discovering the law of falling bodies by close to one hundred years.

2. This idea is, of course, still current in Christian theology. Did not Christ say "I am the Alpha and the Omega"?

3. T. Burnet, *Sacred Theory of the Earth,* p. 249.

4. The percentage of different radioactive elements present in ancient rocks allows the age of their formation to be known with precision. Thorium 232 has a half-life of 14.1 billion years; uranium 238, 4.51 billion years; and uranium 235, 0.71 billion years. Potassium 40 (half-life 1.3 billion years) is used to date ancient fossils. Carbon 14, although it has a relatively brief half-life (5,730 years), is continually formed in the upper atmosphere by cosmic radiation and allows us fairly precisely to date the cessation of exchanges between living organisms and the atmosphere (i.e., their deaths) if it does not go back more than about one hundred thousand years. In chapter 12 I will undertake an analysis of radioactivity as a typically temporal physical phenomenon.

Chapter 6: Time and Space

1. J.-M. Guyau, *Guyau and the Idea of Time,* trans. John A. Michon, Viviane Pouthas, and Constance Greenbaum (New York: North-Holland Publishing, 1988), p. 111.

2. P. Janet, *L'Evolution de la mémoire et de la notion de temps* (The evolution of memory and the notion of time) (Paris: A. Chahine, 1928), p. 182.

3. P. Fraisse, *The Psychology of Time,* p. 203.

4. Piaget, *Child's Conception of Time,* p. 286. Jean Piaget and his team devoted many other studies to the development of the child's notion of time, but these investigated later stages in which reasoning appears, not early infancy.

5. See J. Piaget, *The Construction of Reality in the Child,* p. 44.

6. In the Middle Ages this practice fell into disuse. The geographers of the period drew maps by measuring distances and azimuths between marked points, a method that gave them their peculiar look of spider webs. The invention of the compass and the growing boldness of marine explorers necessitated dispensing with surveys and logs and returning to the direct determination of latitudes and longitudes.

7. Cited in S. Guye and H. Michel, *Time and Space: Measuring Instruments from the Fifteenth to the Nineteenth Century* (New York: Praeger, 1971), p. 141.

8. The interval of time that separates the reception on a ship or a plane of two signals (two blips) emitted at the same instant UTC (coordinated universal time) by two stations of known position allows the difference in the distance between the receptor and the transmitting stations to be calculated. The locus of such a difference is a hyperbola. By repeating the operation with one or several pairs of transmitters, the exact site of the receiving station can be determined. The system Ω, for example, uses three transmitters 5,000 to 6,000 miles away. Comparing the

transmission phase allows the precision of the location to be taken to 0.5 or 1 mile. The Navstar system, or GPS (Global Positioning System), today in development, will use, when completed, a network of eighteen satellites covering the entire Earth, each equipped with synchronized atomic clocks. It will allow a precision of location to within a few meters.

CHAPTER 7: THE SPEED OF LIGHT IS FINITE

1. In fact, the interval between two observations made on the Earth should correspond to the real time between two successive eclipses of Io increased by the difference in the time taken by light to traverse the space between Jupiter and the two successive positions of the Earth.

2. This phenomenon of the composition of velocities is very banal and familiar. For example, it causes a runner to perceive drops of rain as falling not vertically but more and more obliquely, in relation to the runner's own horizontal velocity. It also makes the direction of the wind seem different on a fast sailing ship from the direction it takes on solid ground.

CHAPTER 8: EINSTEIN AND THE THEORY OF RELATIVITY

1. A. Pais, *"Subtle Is the Lord . . .": The Science and Life of Albert Einstein* (New York: Oxford University Press, 1982), p. 139.

2. The experiments of Michelson and Morley were aimed at proving, by interference between two light beams propagating at right angles, a difference of optical path length between the two beams. According to Newtonian mechanics, such a difference of optical path must necessarily be introduced between the two arms of the interferometer as long as it is not at absolute rest.

3. The formulas of transformation of coordinates in special relativity bear Lorentz's name because he had already proposed them, but on an "ad hoc" basis and not by deriving them from the postulates of relativity. In 1905 Einstein was not aware of this work by Lorentz.

4. The famous relationship $E = mc^2$, which established that all mass at rest is a gigantic reservoir of energy, and conversely that all energy has a mass susceptible to universal attraction, is another consequence of the Lorentz transformations that Einstein demonstrated in 1907.

5. Hermann Minkowski, "Space and Time," conference paper given at Cologne in 1908, reprinted in *Annalen der Physik* 47 (1915): 927. The sentence quoted may truly appear excessive. The notion of proper time marking the evolution of a system observed in a reference frame in relation to which this system is at rest suffices to emphasize all the specificity of time in relation to space.

6. The invariant interval between two event-points is given by the formula $d^2 = x^2 + y^2 + z^2 - c^2t^2$. This definition generalizes, in space-time at four dimensions, the notion of spatial distance in Cartesian three-dimensional space.

7. Cited in Ronald W. Clark, *Einstein, the Life and Times* (New York: World, 1971), p. 113.

8. See, for example, B. d'Espagnat, *Reality and the Physicist.*

9. H. Reichenbach, *The Philosophy of Space and Time,* p. 147.
10. H. Reichenbach, *The Direction of Time,* p. 216.

CHAPTER 9: TIME AND THE DANCE OF THE GALAXIES

1. E. Mach, *Die Mechanik in ihrer Entwicklung historich-kritisch Dargestellt.* Translated in English as *The Science of Mechanics.*
2. For example, the forces seen when starting a car or in sharp turns. In the latter case, when implicitly using a frame of reference linked to the car's interior space, it is "centrifugal force" that pushes the passengers into their seats.
3. Einstein himself later recounted how this idea—"the most felicitous of my life"—came to him suddenly, probably one day in November 1907. "I was sitting on a chair at my desk at the office of patents in Bern when a sudden thought came to mind: if a person falls in free fall, he would not feel his own weight! I was staggered. This simple idea made a profound impression upon me. It pushed me towards a relativist theory of gravitation (Pais, *"Subtle Is the Lord,"* p. 179).
4. Einstein assimilated the techniques in question rapidly and well but without feeling an immoderate love for the mathematics per se. He was no doubt thinking about this time in his life when, responding in jest to a little girl who had told him about her troubles with mathematics, he said, "Believe me, mine are just as great!"
5. A. Einstein, *The Meaning of Relativity,* p. 39.
6. The Andromeda galaxy is today accepted to be 2.16 million light-years away.
7. By one of those ironies well known to science, Lemaître became obstinate in his defense of the cosmological constant because he thought it allowed a fundamental difficulty in the model of the initial explosion to be surmounted. This difficulty was that Hubble's time, about 2 billion years, was much less than the age of the Earth, already reliably measured thanks to the natural radioactivity of rocks. These measurements proved that the Earth is more than 3 billion years old (the best contemporary measurement gives 4.55 billion years). What Lemaître did not suspect was that the estimates Hubble made of the distances of the galaxies were greatly underestimated, as just seen in the case of the galaxy of Andromeda, so that the cosmological time characteristic today accepted is much longer, about 15 billion years.
8. See, for example, A. Bouquet, "L'Inflation de l'univers" (The inflation of the universe), *La Recherche* 17, no. 176 (1986): 448–56; S. Hawking, *A Brief History of Time,* pp. 160–68.
9. The adiabatic cooling of a fluid in expansion is a very common phenomenon. It is evident for example in observing that the gas in a bottle of compressed butane cools itself when freely escaping.
10. Due to the almost total symmetry that exists between matter and antimatter, it would then be possible that some of the galactic clusters that exist today are made of antimatter. The light emitted by possible anti-galaxies would be identical to that emitted by ordinary galaxies, but the cosmic radiation coming from these clusters could allow their natures to be confirmed.
11. To the question of the reality or the appearance of expansion, one response

sometimes is that the galaxies are not "truly" fleeing us, that they are not endowed with a true velocity, but that it is the space between them and us that is growing. This idea apparently conforms with the theory of relativity, which makes the metric a function of space and time. But in fact such a formulation distances itself from the spirit of relativity as Einstein conceived it. Remember that there is no way to decide, by physical experiments internal to a system, if this system is at rest or in uniform motion in relation to any exterior reference point. This principle must be applied even if the system in question is a galaxy! Speaking of the "true motion of the galaxies," in opposition to an "apparent motion," therefore makes no sense. Saying that they are fleeing us, or that their distance from us is growing, comes down to a question of semantics.

CHAPTER 10: THE PARADOX OF THE TWINS AND THE STORY OF THE TIME MACHINE

1. P. Langevin, *La Physique depuis vingt ans* (The last twenty years in physics) (Paris: Doin, 1923).

2. M. Sachs, "A Resolution of the Clock Paradox," *Physics Today* 24 (1971): 23.

3. The predicted time difference was 275 ± 20 billionths of a second for the voyage toward the west.

4. J. C. Hafele, "Relativistic Behaviour of Moving Terrestrial Clocks," *Nature* 227 (1970): 270–71; J. C. Hafele, "Relativistic Time for Terrestrial Circumnavigations," *American Journal of Physics* 40 (1972): 81; J. C. Hafele and R. Keating, "Around-the-World Atomic Clocks: Predicted Relativistic Time Gains," *Science* 177 (1972): 166–68.

CHAPTER 11: THE LIMITS OF CAUSAL TIME

1. Laplace had already understood that absolute determinism destroys the orientation of time. In his *Essai philosophique sur les probabilités* (1814; translated in English by F. W. Truscott and F. L. Emory as *A Philosophical Essay on Probabilities* [New York: Dover, 1952]), he noted that for a being intelligent enough to know at any given instant all the forces in nature and all the positions and relative velocities of the particles of which the universe is composed, "nothing would be uncertain, and the future like the past would be there to be seen" (p. 4). The negation of time's objectivity in the sense in which I evoke it here, linked to the absolute determinism of physical theory, was undoubtedly taken to its highest level by Hermann Minkowski. Speaking in 1908 about the implications of relativity theory (before the development of modern cosmogony), he declared that it would be better to think of the world as actually deployed in four dimensions rather than as a world of three dimensions in the process of becoming.

2. J. Monod, *Chance and Necessity,* p. 180.

3. Einstein, it is true, did not often explain himself on this aspect of his epistemological program. The names of Minkowski, on mathematical physics, and of Reichenbach, on epistemology, come to mind first. Einstein insisted more readily

on the principle of relativity and, at a certain period of his life, on Mach's principle. With Einstein, the knowing mind travels from the concept of causality to the principle of reality, rather than the reverse. The appreciation he had for the genius of Newton, whom he saluted as "he who conceived of the fundamental notions of mechanics and physical causality" (letter to the Royal Society in the occasion of the Newton bicentenary, *Nature* 119 [1927]: 467), and the debates he had with Ritz on the primacy of causality in its relationship with entropy are proof of that.

4. In this case, the term of "principle of causality" is imprecise, for it runs the risk of being confused with relativistic principles. It is actually a question of giving precedence to the solutions corresponding to an *initial* causality, for which the effects follow the causes in time and ruling out a *final* causality, for which the effects would precede the causes in time.

5. We must specify *toward the future* or *toward the past,* because the universe we observe with our instruments does not correspond in the least to that which could be called its "present state." The light we receive from distant galaxies testifies to their brightness millions, perhaps even billions, of years ago. Inversely, the photons or the radio messages we send toward extragalactic space will not reach their eventual targets before thousands, millions, or billions of years have passed.

6. R. B. Partridge, "Absorber Theory of Radiation," p. 263.

7. J. G. Cramer, "The Arrow of Electromagnetic Time and the Generalized Absorber Theory," *Foundations of Physics* 13 (1983): 887.

8. A. Einstein, B. Podolsky, and N. Rosen, "Can Quantum-Mechanical Description of Physical Reality Be Considered Complete?" pp. 777–80.

9. A. Aspect et al., "Experimental Tests of Realistic Local Theories via Bell's Theorem," *Physical Review Letters* 47 (1981): 460–63; A. Aspect et al., "Experimental Realization of Einstein-Podolsky-Rosen-Bohm, Gedanken Experiment: A New Violation of Bell's Inequalities," *Physical Review Letters* 49 (1982): 91–96. See also F. Laloe, "Les Surprenantes prédictions de la mécanique quantique" (The surprising predictions of quantum mechanics), *La Recherche* 17, no. 182 (1986): 1358.

10. See the description of this experiment in A. Shimony, "The Reality of the Quantum World," *Scientific American* 258 (Jan. 1988): 36.

11. The interpretation of Wheeler's experiment according to the criteria of reality proposed by Einstein, Podolski, and Rosen can be described as follows. If there is a separate quantum system in each branch of the interferometer and if acting on the first system (by turning the Pockels cell switch on or off) without acting in any way on the second leads the latter to manifest itself as either a wave or a particle, then one must logically conclude that these two properties belong at one and the same time to this system. Quantum mechanics denies this possibility and affirms, on the contrary, that the property of being a wave is incompatible with the property of being a particle. The authors conclude, therefore, that quantum mechanics is incomplete. The resolution of the paradox involves affirming that there is in the interferometer only one single system, an inseparable photon, even if it occupies both branches at one time.

12. As expressed by David Bohm, the finite character of Planck's constant obliges us to replace the pinpoint representation of classical systems in six-dimensional "phase-space"—three dimensions for position and three dimensions for the

impulsion of the particle—by a cell of finite and fixed volume (h^3), whose form depends explicitly on the type of apparatus used. On this subject see D. Bohm, "On Bohr's Views Concerning the Quantum Theory," *Quantum Theory and Beyond* (New York: Cambridge University Press, 1971), p. 33.

13. B. d'Espagnat, "Use of Inequalities for the Experimental Test of a General Conception of the Foundations of Microphysics," *Physical Review,* series D, 11 (1975): 1424–35; B. d'Espagnat, "Use of Inequalities for the Experimental Test of a General Conception of the Foundations of Microphysics," *Physical Review,* series D, 18 (1977): 349–58.

14. B. d'Espagnat, *Les Implications conceptuelles de la physique quantique* (Conceptual implications of quantum physics), *Journal de Physique,* series C2, supplement (1981): 104.

15. D. Bohm, "Foundations of Quantum Mechanics," *Proceedings of the International School of Physics Enrico Fermi,* course 49 (San Diego: Academic Press, 1972).

16. See, for example, I. Prigogine and I. Stengers, *Order out of Chaos;* I. Prigogine, *From Being to Becoming.*

17. O. Costa de Beauregard, *Time, the Physical Magnitude.*

18. On this subject see W. Salmon, "Probabilistic Causality," *Pacific Philosophy Quarterly* 61 (1980): 50–74; R. A. Healey, "Temporal and Causal Asymmetry," in *Space, Time, and Causality,* ed. R. Swinburne (Boston: D. Reidel, 1983), p. 79–103; and the position of Costa de Beauregard in d'Espagnat, *Reality and the Physicist,* pp. 229–31.

CHAPTER 12: THE SUSPENDED TIME OF THE ATOMS

1. E. Rutherford and F. Soddy's articles summarizing their observations ("Condensation of Radioactive Emanations," *Philosophy Magazine,* May 1903, 561–76; "Radioactive Change," ibid., pp. 576–91) do, however, contain penetrating remarks on the nature of radioactivity and its possible consequences: "The energy of radioactive change must therefore be at least twenty-thousand times, and may be a million times, as great as the energy of any molecular change. . . . Hence there is no reason to assume that this enormous store of energy is possessed by the radioelements alone. It seems probable that atomic energy in general is of a similar, high order of magnitude, although the absence of change prevents its existence being manifested. . . . The maintenance of solar energy, for example, no longer presents any fundamental difficulty if the internal energy of the component elements is considered to be available, i.e., if processes of sub-atomic change are going on" (590).

2. The initial population of atoms of a given type, homogeneous and thus describable with a minimum of information, is transformed into a heterogeneous population of atoms of different types. The description of their distribution in the sample would require more detailed information. See R. Lestienne, "Caractères de la durée physique: II Mesure du temps" (Characteristics of physical duration: Measuring Time), *Scientia* 107 (1972): 278–306.

3. See, for example, P. L. Knight and P. W. Milonni, "Long-Time Deviations

from Exponential Decay in Atomic Spontaneous Emission Theory," *Physics Letters* 56A4 (1976): 275.

4. See Prigogine, *From Being to Becoming,* p. 188.

5. For example, such an individual sees clocks uncounting hours, rivers returning to their sources, etc. From this perspective the world has the same appearance as it would for us when watching a film wound backwards and starting at the end.

CHAPTER 13: THE AGE OF THINGS, ORDER AND CHAOS,
ENTROPY AND INFORMATION

1. It changes form if t is changed to $-t;$ in other words, it is constructed to apply only to predictions toward the future and does not give correct results if applied in reverse.

2. It is not a question here of the particular example of K° meson decay or of the possible violation of the time-reversal symmetry by nuclear forces considered in the preceding chapter. It is difficult to imagine any link between this particular property of elementary particles and the general law of the dissipation of heat.

3. A major objection that might come to the minds of thermodynamic specialists concerns the many causes of irreversibility, in particular the diffusion of matter. But it must be observed that the diffusion of matter in an isothermic environment can be considered as a particular case of isothermic expansion of a fluid endowed with an appropriate partial pressure, a consumer of heat.

4. The truth is that Mayer did indeed articulate the idea of the equivalence between heat and mechanical energy well before Joule, but it was Joule who produced the experimental proof of this equivalence by measuring the mechanical equivalent of the calorie.

5. A bit later, in the years 1870–90, a true mystique of energy would develop under the name of "energetism." One of its most ardent promoters, W. Ostwald, was the first to suggest, in 1892, a unity of nature between matter and energy. Energy, he argued, is the only true entity in the world and matter is not the carrier, but rather the manifestation of energy.

6. $\int dQ/T,$ where dQ measures an elementary exchange of heat between the system and the exterior surroundings and $T,$ the absolute temperature at which this exchange takes place, the integral being taken from an initial state taken as reference until the state considered, by means of an ideal reversible procedure.

7. H. Bergson, *Creative Evolution,* p. 243.

8. See, for example, I. Prigogine and I. Stengers, *Entre le temps et l'éternité,* pp. 117, 184.

9. To perform this enumeration, Boltzmann proposed dividing the total of the possible values for the positions and the velocities of each gas molecule into small segments of finite size and counting the number of the distinct configurations that give the same global appearance to the volume of gas (local volume, pressure, and temperature).

10. L. Boltzmann, *Wissenschaftliche Abhandlungen,* vol. 2, p. 121.

11. Ibid., vol. 3, p. 567

12. A mole contains Avogadro's number (6×10^{23}) of real molecules. In nor-

mal conditions of temperature and pressure, a mole of perfect gas occupies 22.4 liters and thus constitutes a useful macroscopic unit.

13. Boltzmann wrote: $S = k \ln W$. The epistemological import of this formula appears so great that, graven in stone, it serves as an epitaph on the tomb of its author. In this expression, S is entropy, k a constant (since called Boltzmann's constant), and W is the number of different ways in which the macroscopic properties of a system can be reproduced, giving different values to its microscopic degrees of freedom—for example, to the position and velocity of each of the particles of which it is composed. Today, in accord with Gibbs's work, this result would be described by saying that the entropy of a system of totally fixed energy is proportional to the logarithm of the volume of phase-space compatible with the system's known macroscopic properties.

14. R. Feynman, *The Feynman Lectures on Physics,* vol. 1, pp. 52–53.

15. H. Poincaré, *Science et méthode* (Paris, 1908; rpt. Paris: Gauthiers-Villars, 1956), p. 66; Poincaré, *Science and Method* (New York: Dover, 1952), p. 66.

16. A system composed of n independent material particles possesses $6n$ degrees of freedom, since to characterize the state of the system it is necessary to fix the value of the three position components and the three velocity components of each of these particles.

17. Camille Flammarion's book *L'Astronomie populaire* (Popular astronomy), published in 1880, ends with this sublime exhortation: "Let us stand upright before the heavens and have henceforth only one and the same banner: Progress through Science!"

18. P. T. Landsberg, "Time in Statistical Physics," p. 1108; Landsberg, "Usage et limites du concept d'entropie" (The use and limits of entropy), *Communications* 41 (1985): 63.

19. P. Lecomte du Nouy, *L'Homme devant la science,* p. 49.

20. R. Lestienne, "Entropie, temps mécanique et flèche cosmologique," 337–58.

21. E. Borel, *Le Hasard* (Chance) (Paris: Alcan, 1914), p. 307.

22. L. Boltzmann, *Theoretical Physics and Philosophical Problems,* p. 24.

23. D. Bohm, *Foundations of Quantum Mechanics* (San Diego: Academic Press, 1971), p. 412.

24. A. Eddington, *New Pathways in Science,* p. 68.

25. Without a change in entropy.

26. R. Lestienne, "A la mémoire de Ludvig Boltzmann," p. 3.

CHAPTER 14: DISSIPATIVE STRUCTURES, CYCLIC REACTIONS

1. Despite, in addition to, or because of his ideas on thermal disorder, Boltzmann was an ardent apostle of Darwinism.

2. J. Tonnelat, in *Thermodynamique et biologie* (Thermodynamics and biology), vol. 2 (Paris: Malaine, 1978), p. 16, insists on distinguishing spatial disorder, which lacks a direct connection with entropy, and energetic disorder, which is linked to entropy through the concept of complexity.

3. Prigogine, *From Being to Becoming,* p. 109

4. A. Pacault, P. de Kepper, and P. Hanusse, "Sustained Chemical Dissipative Systems: Illustration of Thermodynamic Time," *Comptes rendus de l'Académie des Sciences* 280B (1975): 157–61

5. See, for example, I. Epstein et al., "Oscillating Chemical Reactions," *Scientific American* 249 (Mar. 1983): 112–23.

6. These are open systems, because only a permanent flux of matter and energy allows the system to achieve and maintain the considered state of affairs. In the autocatalytic stages, certain molecules produced allow their own formation to be multiplied; in the cross-catalytic stages, certain molecules are necessary to produce other molecules, which produce in their turn molecules of the first type.

7. I. Prigogine and I. Stengers, "La nouvelle alliance," p. 300.

Chapter 15: From the Time of Things to the Time of Living Beings

1. See I. Prigogine, "La Thermodynamique de la vie" (Thermodynamics and life), *La Recherche* 3, no. 24 (1972).

2. Mitochondria also make ATP but in another way, by directly oxidizing (in the presence of oxygen) glucides.

3. A. T. Winfree, *The Geometry of Biological Time,* p. 285.

4. In procaryotes, phosphofructokinase is activated directly by ADP itself. Other activators and inhibitors can also play a role in the control of the activity of this enzyme. See A. Sols et al., *Metabolic Interconversion of Enzymes 1980* (New York: Springer-Verlag, 1981), p. 111; A. Sols, *Current Topics in Cellular Regulation* (San Diego: Academic Press, 1981), vol. 18, pp. 77–101.

5. H. Maturana and F. Varela, *Autopoiesis and Cognition: The Realization of the Living* (Dordrecht, Holland: D. Reidel, 1980), p. 88.

6. Cited by J.-P. Dupuy, "Histoires de Cybernétiques" (History of cybernetics), *Cahiers du CREA,* no. 7, Ecole Polytechnique, 1985, p. 83.

Chapter 16: What Is Life?

1. Cited by B. L. Strehler, *Time, Cells, and Aging,* p. 37.

2. J. Monod, *Chance and Necessity.*

3. This observation linking life to a specific activity such as reproduction or simply metabolism can be made for all living organisms in a state of hibernation or in a cryogenized state. Can one, must one, call them truly living? Shouldn't they rather be considered to be in a state of "suspended animation"?

4. H. Atlan, *Entre le cristal et la fumée,* p. 14.

5. P. Teilhard de Chardin, *The Phenomenon of Man.*

6. F. Jacob, *The Logic of Life,* p. 307.

7. See, for example, the proceedings of the symposium entitled "Mathematical Challenges to the Neo-Darwinian Interpretation of Evolution," edited by the Wistar Institute Press, Philadelphia, 1967, a symposium at which some of the greatest names in science participated.

8. In France, one of the best known of these is Professor Grassé. See, for example, his interview "L'Evolution sans projet" (Evolution without purpose) in *Le Darwinisme aujourd'hui* (Darwinism today) (Paris: Seuil, 1979), p. 129.

9. F. Jacob, in *Le Darwinisme aujourd'hui*, p. 154.

10. J. Campbell, *The Grammatical Man.*

11. H. Von Foerster, "On Self-Organizing Systems and Their Environments," in *Self-Organizing Systems,* ed. H. Young and S. Cameron (Elmsford, N.Y.: Pergamon, 1960), pp. 31–50.

12. B. L. Strehler, "Genetic Instability as the Primary Cause of Aging," *Experimental Gerontology* 21 (1986): 283–319.

13. W. R. Ashby, "Principles of the Self-Organizing Systems," in *Principles of Self-Organization,* ed. H. Von Foerster and G. Zopf (Elmsford, N.Y.: Pergamon, 1962), pp. 255–78.

14. H. Atlan, *L'Organisation biologique,* p. 251.

15. A. Danchin, *La Recherche* 19, no. 201 (1988): 878. See also, from the same author, *L'Oeuf et la poule: Histoires du code génétique* (The egg and the chicken: stories of the genetic code) (Paris: Fayard, 1983).

16. See S. L. Miller and L. E. Orgel, *The Origin of Life on the Earth* (New York: Prentice-Hall, 1972).

17. *Encyclopédie Alpha des Sciences et Techniques,* vol. 1 (La Grange-Batelière, 1976), p. 7.

CHAPTER 17: LIFE

1. A. Reinberg, *Des rythmes biologiques à la chronobiologie* (Biological rhythms in chronobiology) (Paris: Gauthiers-Villars, 1977), p. 11. Reinberg specifies that this "discovery" was in fact a rediscovery, since observations of this type had already been made in the nineteenth century. They were practically overlooked at that time.

2. J. W. Hastings and B. M. Sweeney, "A Persistent Diurnal Rhythm of Luminescence in Gonyaulax Polyedra," *Biological Bulletin* 115 (1958): 440–58.

3. M. B. Wilkins, "Circadian Rhythms in Plants," in *Biological Aspects of Circadian Rhythms,* ed. J. N. Mills (New York: Plenum, 1973).

4. The axon of certain human nerve cells measures more than a meter.

5. More precisely, it is the difference in electrical potential between the interior and the exterior of the neuron, on both sides of the membrane wall, caused by the differential concentration in ions of different types in these two environments. At a state of rest, the external milieu is much richer in Na^+, Ca^{++}, and Cl^-, and the internal milieu is much richer in K^+ ions.

6. Dendrites are the root-shaped appendages of the neurons. They receive incoming nervous information and transmit it to the cell body.

7. Quiang Yu et al., "Behaviour Modification by *in Vitro* Mutagenesis of a Variable Region within the *Period* Gene of *Drosophila*," *Nature* 326 (1987): 765–69.

8. See J. D. Vincent, *Biologie des Passions* (The biology of emotion) (Paris: Editions Odile Jacob, 1986), p. 314.

CHAPTER 18: THE TEMPORAL CONTROL OF DEVELOPMENT

1. A human chromosome, whose size is no more than the hundredth of a millimeter, easily contains more than ten linear centimeters of DNA! It is important

to note also that in addition to being coiled, DNA is often twisted, like a string too tightly wound that has wrapped around itself. This twisting normally prevents the separation of the two strands of the helix; at the moment of replication, it is released thanks to the action of a specific enzyme.

2. This particularity explains Chargaff's observation. The adenine carried by a strand of the double helix associates itself exclusively with the thymine carried by the other strand, and the guanine associates with the cytosine. These exclusive associations are made possible through the complementary forms of these molecules, which allow the establishment of "hydrogen bonds" from one to the other.

3. The portion of the gene placed at its beginning and controlling its expression is called the "promoter." If the promoter is active, the gene is expressed, that is, transcribed into messenger RNA (later this messenger RNA can be translated into precisely the sort of protein coded by the gene). An inactive promoter blocks the transcription stage.

4. W. J. Gehring, "On the Homeobox and Its Significance," *Bioessays* 5 (1986): 3–4.

5. The homology at the level of the protein can be greater than that observed at the level of the DNA because the same amino acid can sometimes be coded by different triplets of nucleic bases.

6. See M. Blanc, "Les Races humaines existent-elles?" (Do human races exist?), *La Recherche*, 13, no. 135 (1982): 930.

7. M.-C. King and A. C. Wilson, "Evolution at Two Levels," p. 107.

8. J. P. Changeux, *Neuronal Man,* p. 249.

9. A. G. Knudson, "Genetics of Human Cancer," p. 231.

CHAPTER 19: THE BRAIN, THE MIND, AND TIME

1. The molecules of neurotransmitters, whose function is to transmit information from one neuron to another by way of the synaptic space, are liberated by vesicules situated on the "boutons" (buttons) of the upstream cell's axonal arborescences. They do not really penetrate the downstream cell but fix themselves to the wall of the latter's synaptic membrane, which temporarily modifies the permeability of certain ionic channels and thus creates an electrical signal modifying the membrane's potential.

2. A. M. Turing, "On Computable Numbers, with an Application to the Entscheidungs Problem," *Proceedings of the London Mathematical Society* 4, no. 2 (1937): 230–65.

3. W. S. McCulloch and W. Pitts, "A Logical Calculus of the Ideas Immanent in Nervous Activity," *Bulletin of Mathematic Biophysics* 5 (1943): 115–33.

4. This succession represents the decomposition of the number considered into increasing powers of two. For example, in binary writing, the number 10 is written 1010, since $10 = (0 \times 2^0) + (1 \times 2^1) + (0 \times 2^2) + (1 \times 2^3)$. This decomposition is unique.

5. The redundancy of the information stored in the brain does not necessarily mean a diminution of the quantity of memorizable information. In exchange for the existence of multiple copies of one piece of information, several different pieces of information can be stored in the same synaptic groups.

6. S. Thorpe, "Identification of Rapidly Presented Images by the Human Visual System," *Perception* 17 (1988): 415 (A77).

7. F. J. Varela et al., "Perceptual Framing and Cortical Alpha Rhythms," *Neurophsychologia* 19 (1981): 675–86.

8. See, for example, "Les Nouveaux Ordinateurs" (The new computers), *La Recherche* (special issue), 18, no. 204 (Nov. 1988).

9. C. Von der Malsburg and E. Bienenstock, "A Neural Network for the Retrieval of Superimposed Connection Patterns," *Europhysics Letters* 3 (1987): 1243–49; Bienenstock and Von der Malsburg, "A Neural Network for Invariant Pattern Recognition," *Europhysics Letters* 4 (1987): 121–26.

10. A. Pellionisz and R. R. Llinas, "Space-Time Representation in the Brain: The Cerebellum as a Predictive Space-Time Metric Tensor," *Neuroscience* 7 (1982): 2249–70; Pellionisz and Llinas, "Tensor Network Theory of the Metaorganization of Functional Geometries in the Central Nervous System," *Neuroscience* 16 (1985): 245–73.

11. Of all the capacity, in the sense that the brain stores in its memory an unsuspected amount of detail, which can be brought out at need, as verified so often in criminal trials in the courts of law; of all the fidelity, because it is capable of storing, year after year, a considerable quantity of information that is later saved without distortion. It is currently being verified that certain events long held in memory are difficult or slow to come back, but details are just as clear. In a purely spatial (network) model of memory, the events to be memorized are, so to speak, stacked one on the other in a latent state in an entire ensemble of nerve cells with their synapses. At each new fact inscribed, the synaptic weights—that is, the efficiency of the synapses—are changed, even imperceptibly. It is difficult to understand how the exact nature of the memories previously written remains unchanged.

12. According to very recent measurements, the quantity of information transported in a single nerve fiber, measured according to the norms defined by Shannon's theory of information, would be at least on the order of 40 percent higher than the information transported by the single average frequency of discharge in the optical nerve, 90 percent higher than that in the lateral geniculate body and in the primary visual cortex, and around 125 percent higher in the temporal cortex, the usual final stage of the task of pattern recognition. See B. J. Richmond et al., "Lateral Geniculate Nucleus Neurons in Awake Behaving Primates," *Society for Neuroscience Abstracts* 14 (1988): 309.

13. B. L. Strehler and R. Lestienne, "Evidence on Precise Time-Coded Symbols and Memory of Patterns in Monkey Cortical Neuronal Spike Trains," *Proceedings of the National Academy of Science USA* 83 (1986): 9812–16; Lestienne and Strehler, "Time Structure and Stimulus Dependence of Precisely Replicating Patterns Present in Monkey Cortical Neuronal Spike Trains," *Brain Research* 437 (1987): 214–38.

14. C. M. Gray and W. Singer, "Stimulus Specific Neuronal Oscillations in the Cat Visual Cortex: A Cortical Functional Unit," *Society for Neuroscience Abstracts* 13 (1987): 1449; R. Eckhorn et al., "Coherent Oscillations: A Mechanism of Feature Linking in the Visual Cortex?" *Biological Cybernetics* 60 (1988): 121–30.

15. H. Bergson, *Time and Free Will,* p. 100.

16. W. James, *Psychology: The Briefer Course* (Henry Holt, 1892), p. 284.

17. Ibid., p. 247

18. In a letter addressed to the family of his friend M. Besso on the occasion of the death, A. Einstein wrote: "For us believing physicists the distinction between past, present, and future is only an illusion, even if a stubborn one." Cited by B. Hoffmann in *Albert Einstein, Creator and Rebel* (New York: Viking Press, 1972), p. 258.

19. See, for example, A. Duval, "Analyse spectrale de la notion du temps: La non-univocité du temps" (Spectral analysis of the notion of time; the equivocality of time), *Revue des Sciences Philosophiques et Théologiques* 68 (1984): 513–46.

20. H. Bergson, *Matter and Memory*, p. 30.

21. M. Heidegger, in M. Haar, *Martin Heidegger* (Paris: Ed. de l'Herne, 1983), p. 33.

Epilogue

1. This does not signify that the exploration has been completed on this point or that the sense we give today to these concepts is not susceptible to new refinements; indeed, the truth is quite the contrary. But that is the entire spirit of the scientific enterprise. More than the art of dominating nature and harnessing forces so as to use them to our profit through technology, science is first the art of *progressively adapting our language to nature.*

2. Despite the summary character of this outline, note that the theory of bifurcations in the thermodynamics of irreversible processes already lets us understand how the future is not always certain, as long as the road taken depends on the properties of the environment or of the universal chaos that do not allow for any prediction.

3. F. Hegel, *Hegel's Philosophy of Nature*, p. 37.

4. H. Barreau, "Temps et devenir"; quotation on p. 17.

5. The second principle of thermodynamics affirms that all concrete physical systems are sources of entropy.

6. H. Bergson, *The Creative Mind*, p. 341.

7. H. Bergson, *Creative Evolution*, p. 224.

Bibliography

Abbott, L. "The Mystery of the Cosmological Constant." *Scientific American* 258 (1988): 82.

Altarev, I. S., et al. "A New Upper Limit on the Electric Dipole Moment of the Neutron." *Physics Letters* 102B (1981): 13–16.

Atlan, H. *Entre le cristal et la fumée* (Between the crystal and the smoke). Paris: Seuil, 1979.

―――. *L'Organisation biologique et la théorie de l'information* (Biological organization and information theory). Paris: Hermann, 1972.

Attali, J. *Histoires du temps* (Stories of time). Paris: Fayard, 1982.

Audouze, J. "L'Expansion de l'univers sera-t-elle éternelle?" (Is the expansion of the universe eternal?) *La Recherche* 13, no. 136 (1982): 1080.

Augustine, Saint. *The City of God.* Vol 11. Trans. Marcus Dods. New York: Modern Library, 1950.

Bachelard, G. *La Dialectique de la durée* (Dialectic of duration). Paris: P.U.F., 1936.

―――. *L'Intuition de l'instant* (Grasping the Instant). Paris: Stock, 1932.

Barreau, H. "Bergson et la théorie de la relativité" (Bergson and the theory of relativity). *Cahiers fundamenta scientiae* (University of Strasbourg) 4 (1974): 1–46.

―――. "Temps et devenir" (Time and becoming). *Rev. Phil. Louvain* 86 (1988): 5–36.

―――, ed. *Temps de la vie et temps vécu* (Life time and lived time). Paris: Editions du CNRS, 1982.

Bergson, H. *Creative Evolution.* Trans. Arthur Mitchell. Lanham, Md.: University Press of America, 1984.

―――. *The Creative Mind.* Trans. M. L. Andison. New York: Philosophical Library, 1986.

―――. *Duration and Simultaneity, with Reference to Einstein's Theory.* Trans. Leon Jacobson. Indianapolis: Bobbs-Merrill, 1965.

―――. *Matter and Memory.* Trans. Nancy M. Paul and W. Scott Palmer. New York: Zone, 1988.

―――. *Time and Free Will: An Essay on the Immediate Data of Consciousness.* Trans. F. L. Pogson. London: G. Allen & Co. 1950.

Black, I. B., et al. "Biochemistry of Information Storage in the Nervous System." *Science* 236 (1987): 1263–68.

Boissin, J., and B. Canguilhem. "Les Rythmes circannuels chez les mammifères" (Circannual rhythyms among mammals). *Archives Internationales de Physiologie et de Biochimie* 96 (1988): A289–A345.

Boltzmann, L. *Theoretical Physics and Philosophical Problems: Selected Writings.* Trans. Paul Foulkes. Boston: D. Reidel, 1974.

———. *Wissenschaftliche Abhandlungen* (Scientific papers). Vol. 2, 1909. New York: Chelsea, 1968.

Brooks, D. R., and E. O. Wiley. *Evolution as Entropy: Towards a Unified Theory of Biology.* Chicago: University of Chicago Press, 1988.

Burnet, T. *Sacred Theory of the Earth.* London: Centaur, 1965.

Cabrera, B. "First Results from a Superconductive Detector for Moving Magnetic Monopoles." *Physical Review Letters* 48 (1982): 1378.

Campbell, J. *The Grammatical Man.* New York: Simon and Schuster, 1982.

Capek, M., ed. *The Concepts of Space and Time: Their Structures and Their Development.* Boston Studies in the Philosophy of Science, vol. 22. Dordrecht, Holland: D. Reidel, 1976.

Cazenave, M., comp. *Création et désordre* (Creation and disorder). Paris: L'Originel, 1987.

Centre International de Synthèse. *Galilée: Aspects de sa vie et de son oeuvre* (Galileo: Aspects of his life and of his work). Paris: P.U.F. 1968.

Chambadal, P. *Evolution et applications du concept d'entropie* (Evolution and applications of the concept of entropy). Paris: Dunod, 1963.

Changeux, J.-P. *Neuronal Man: The Biology of Mind.* New York: Oxford University Press, 1986.

Churchland, P. S. *Neurophilosophy.* Cambridge: MIT Press, 1986.

Clavelin, M. *The Natural Philosophy of Galileo.* Cambridge: MIT Press, 1974.

Cohen-Tannoudji, G., and M. Spiro. *La Matière-espace-temps* (Matter-space-time). Paris: Fayard, 1986.

Coppens, Y. *Le Singe: L'Afrique et l'homme* (The ape: Africa and man). Paris: Fayard, 1983.

Costa de Beauregard, O. *Le Second principe de la science du temps* (The second principle of the science of time). Paris: Seuil, 1963.

———. *Time, the Physical Magnitude.* Boston: D. Reidel, 1987.

Crombie, A. C. *Augustine to Galileo: The History of Science, A.D. 400–1650.* Cambridge: Harvard University Press, 1953.

Cronin, V. *The View from Planet Earth: Man Looks at the Cosmos.* New York: William Morrow, 1981.

Darwin, C. *On the Origin of Species.* Cambridge: Harvard University Press, 1964.

Davies, P. C. W. *The Physics of Time Asymmetry.* Berkeley: University of California Press, 1974.

d'Espagnat, B. *Un Atome de sagesse* (An atom of wisdom). Paris: Seuil, 1982.

———. *Reality and the Physicist: Knowledge, Duration and the Quantum World.* Trans. J. C. Whitehouse and Bernard d'Espagnat. New York: Cambridge University Press, 1989.

———. *In Search of Reality.* New York: Springer-Verlag, 1983.

Duhem, P. *Le Système du monde* (The world system). Paris: Hermann, 1965.

Eddington, A. *New Pathways in Science.* New York: Macmillan, 1935.

Einstein, A. "Eloge nécrologique de E. Mach" (In praise of E. Mach on his death). *Physicalische Zeitschrift* 17 (1916): 101–4.

———. *The Meaning of Relativity,* 5th ed. Princeton, N.J.: Princeton University Press, 1950.

———. *Relativity: The Special and General Theory.* New York: H. Holt and Co. 1920.

———, and L. Infeld. *The Evolution of Physics: The Growth of Ideas.* . . . New York: Simon and Schuster, 1938.

———, B. Podolsky, and N. Rosen. "Can Quantum-Mechanical Description of Physical Reality Be Considered Complete?" *Physical Review* 47 (1935): 777–80.

Elkana, Y. "Boltzmann's Scientific Research Programme and Its Alternatives." *Some Aspects of the Interaction between Science and Philosophy.* Jerusalem: Van Leer, 1971.

Ellis, J. "The Very Large and the Very Small." In Mulvay, *The Nature of Matter,* 126–43.

Emery, E. *Temps et musique* (Time and music). Lusanne: L'âge d'Homme, 1975.

Farrington, B. *Science in Antiquity.* New York: Oxford University Press, 1969.

Feynman, R. *The Character of Physical Law.* Cambridge: MIT Press, 1967.

———. *The Feyman Lectures on Physics.* Vol. 1. Reading, Mass.: Addison-Wesley, 1963.

Figuier, L. *Vie des savants illustrée du XVIIIe siècle* ((The lives of famous thinkers of the eighteenth century). Paris: Lacroix, 1870.

Fraisse, P. "Perception and Estimation of Time." *Annual Revue of Psychology* 35 (1984): 1–36.

———. *The Psychology of Time.* New York: Harper & Row, 1963.

———, ed. *Du temps biologique au temps psychologique* (From biological time to psychological time). Symposium of the Association of Scientific Psychologists of the French Language. Poitiers, 1977. Paris: P.U.F. 1979.

Fraser, J. T. *The Voices of Time.* Amherst: University of Massachusetts Press, 1981.

Fraser, J. T., F. C. Haber, and G. H. Muller, eds. *The Study of Time I.* New York: Springer-Verlag, 1972.

Fraser, J. T., and N. Lawrence, eds. *The Study of Time II.* New York: Springer-Verlag, 1975.

Fraser, J. T., N. Lawrence, and D. Park, eds. *The Study of Time III.* New York: Springer-Verlag, 1978.

———, eds. *The Study of Time IV.* New York: Springer-Verlag, 1981.

French, A. P. *Einstein: A Centenary Volume.* Cambridge: Harvard University Press, 1979.

Gal-Or, B. "The Crisis about the Origin of Irreversibility and Time Anisotropy." *Science* 176 (1972): 1117.

Gehring, W. J. "Homeo Boxes in the Study of Development." *Science* 236 (1987): 1245–52.

Gehring, W. J., and Y. Hirm Y. "Homeotic Genes and the Homeobox." *Annual Review of Genetics* 20 (1986): 147–73.

Gold, T., ed. *The Nature of Time*. Ithaca: Cornell University Press, 1967.

Gould, S. J. *Ever since Darwin: Reflections in Natural History*. New York: Norton, 1977.

———. *Time's Arrow, Time's Cycle: Myth and Metaphor in the Discovery of Geological Time*. Cambridge: Harvard University Press, 1987.

Graves, J. C. *The Conceptual Foundations of Contemporary Relativity Theory*. Cambridge: MIT Press, 1971.

Guye, S., and H. Michel. *Mesures du temps et de l'espace* (Measuring time and space). Fribourg: Office du Livre, 1970.

Halberg, F. "Les Rythmes biologiques et leurs mécanismes" (Biological rhythms and their mechanisms). In Fraisse, ed., *Du temps biologique au temps psychologique*, pp. 21–71

Hawking, S. *A Brief History of Time: From the Big Bang to Black Holes*. New York: Bantam, 1988.

Hegel, G. W. F. *Hegel's Philosophy of Nature*. Trans. A. V. Miller. Oxford: Clarendon, 1970.

Heisenberg, W. *Physics and Beyond: Encounters and Conversations*. New York: Harper & Row, 1971.

———. *Physics and Philosophy: The Revolution in Modern Science*. New York: Harper, 1958.

Holton, G. *The Scientific Imagination: Case Studies*. New York: Cambridge University Press, 1978.

Hoppe, E. *Histoire de la physique* (History of physics). Paris: Payot, 1928.

Infeld, L. "On the Structure of Our Universe." In *A. Einstein, Philosopher-Scientist*, p. 477–99. New York: Cambridge University Press, 1970.

Jacob, F. *The Logic of Life: A History of Heredity*. Trans. Betty E. Spillmann. New York: Pantheon, 1974.

———. *The Possible and the Actual*. Seattle: University of Washington Press, 1982.

———. *The Statue Within: An Autobiography*. New York: Basic Books, 1988.

King, M.-C., and A. C. Wilson. "Evolution at Two Levels in Humans and Chimpanzees." *Science* 188 (1975): 107–16.

Kinsey, K. F., et al. "Measurement of the Lifetime of Positive Pions." *Physical Review* 144 (1966): 1132–37.

Knudson, A. G. "Genetics of Human Cancer." *Annual Review of Genetics* 20 (1986): 231–51.

Koyré, A. *From the Closed World to the Infinite Universe*. Baltimore: Johns Hopkins University Press, 1957.

———. *Galileo Studies*. Atlantic Highlands, N.J.: Humanities Press, 1978.

Krauss, L. M. "Dark Matter in the Universe." *Scientific American* 255 (1986): 50.

Laborit, H. *Dieu ne joue pas aux dés* (God does not throw dice). Paris: Grasset, 1987.

Landsberg, P. T., comp. *The Enigma of Time*. Bristol: Adam Hilger, 1982.

———. "Nature: The Beginning and End of the Universe." *Nature, Time, and History* (Nijmegen Studies in Philosophy of Nature and Sciences) 4 (1985): 77–108.

———. "Time in Statistical Physics and Special Relativity." *Studium Generale* 23 (1970): 1108.

————. "Usages et limites du concept d'entropie" (Uses and limitations of the concept of entropy). *Communications* 41 (1985): 63.

Lang, K. R., and O. Guinguish. *A Source Book in Astronomy* (with the original articles of Hubble, Lemaître, and Friedmann). Cambridge: Harvard University Press, 1979.

Langevin, J., and M. Paty. "Le Séjour d'Einstein en France en 1922" (Einstein's 1922 visit to France). *Cahiers Fundamenta Scientiae,* no. 93. University Strasbourg, 1976.

Layzer, D. "The Arrow of Time." *Scientific American* 233 (1975): 56–69.

————. "The Arrow of Time." *The Astrophysical Journal* 206 (1976): 559.

————. "The Strong Cosmological Principle, Indeterminacy, and the Direction of Time." In *The Nature of Time,* ed., J. H. Mulvey, p. 111. Ithaca: Cornell University Press, 1967.

Lecomte du Nouy, P. *L'Homme devant la science* (Man confronts science). Paris: Flammarion, 1969.

Lemaître, G. "The Cosmological Constant." In *A. Einstein, Philosopher-Scientist,* p. 439–456. New York: Cambridge University Press, 1970.

Lestienne, R. "Entropie, temps mécanique et flèche cosmologique" (Entropy, Mechanical Time, and the cosmological arrow). *Scientia* 115 (1980): 337–58.

————. "L'Espace perdu et le temps retrouvé" (Space lost and time recovered). *Communications* 41 (1985): 5–26.

————. "From Physical to Biological Time." *Mechanisms of Ageing & Development* 43 (1988): 189–228.

————. "A la mémoire de Ludvig Boltzmann: L'Entropie est-elle objective?" (In memory of Ludvig Boltzmann: Is entropy objective?) *Fundamenta Scientiae* 8, no. 2 (1987): 173–84.

————. "Unité et ambivalence du concept de temps physique" (Agreement and ambivalence of the concept of physical time). *Cahiers d'histoire et de philosophie des sciences* (CNRS) 9 (1979).

Levy-Leblond, J. M. "L'Espace, le temps et les quantons" (Space, time and quanta). In *L'Espace et le temps aujourd'hui* (Space and time today), ed. Emile Noël. Paris: Seuil, 1983.

Locke, J. *An Essay Concerning Human Understanding.* Ed. P. H. Nidditch. Oxford: Clarendon, 1970.

Longo, M. J. "Massive Magnetic Monopoles: Indirect and Direct Limits on Their Number, Density, and Flux." *Physical Review* D25 (1982): 2399.

Mach, E. *Die Mechanik in ihrer Entwicklung historich-kritisch Dargestellt.* Brockhaus, 1883. Translated in English as *The Science of Mechanics: A Critical and Historical Exposition of Its Principles.* Trans. Thomas J. McCormack. Chicago: Open Court, 1893

Mehra, J. "Einstein, Hilbert, and the Theory of Gravitation." In Mehra, ed., *The Physicist's Conception of Nature,* pp. 92–178.

————. Ed. *The Physicist's Conception of Nature.* Oxford: Clarendon, 1981.

Merleau-Ponty, J. *Cosmologie du XXe siècle: Études épistémologique et historique des théories de la cosmologie contemporaine* (Twentieth-century cosmology: epistemological and historical studies of contemporary cosmologies' theories). Paris: Gallimard, 1965.

————. *La Science de l'univers à l'âge du positivisme* (The sciences of the universe in the age of positivism). Paris: Vrin, 1983.

Milgram, M., and H. Atlan. "Probabilistic Automata as a Model for Epigenesis of Cellular Networks." *Journal of Theoretical Biology* 103 (1983): 523–547.

Miller, A. I. *Albert Einstein's Special Theory of Relativity: Emergence (1905) and Early Interpretation (1905–1922)*. Reading, Mass.: Addison Wesley, 1981.

Minors, D. S., and J. M. Waterhouse. "The Sleep-Wakefulness Rhythms: Exogenous and Endogenous Factors (in Man)." *Experientia* 40 (1984): 410–16.

Misner, C. W., K. S. Thorne, and J. A. Wheeler. *Gravitation.* San Francisco: Freeman, 1973.

Monod, J. *Chance and Necessity: An Essay on the Natural Philosophy of Modern Biology.* Trans. Austryn Wainhouse. New York: Knopf, 1971.

Morin, E. *Science avec conscience* (Conscientious science). Paris: Fayard, 1982.

Muirhead, H. *The Special Theory of Relativity.* New York: Macmillan, 1973.

Mulvey, J. H., ed. *The Nature of Matter.* Oxford: Clarendon, 1981.

Nicolescu, B. *Nous, la particule et le monde* (The particle, the world, and us). Geneva: Le Mail, 1985.

Noël, E. *Le Darwinisme aujourd'hui* (Darwinism today). Paris: Seuil, 1979.

Omnes, R. *L'Univers et ses métamorphoses* (The universe and its metamorphoses). Paris: Hermann, 1973.

Pacault, A., and C. Vidal. *A chacun son temps* (To each his time). Paris: Flammarion, 1975.

Partridge, R. B. "Absorber Theory of Radiation and the Future of the Universe." *Nature* 244 (1973): 263.

Paty, M. "La Critique rationnaliste de la création au XVIIIᵉ siecle" (The twentieth-century rationalist criticism of creation). *Dialectica* 37 (1983): 185–200.

————. "Einstein et la pensée de Newton" (Einstein and the mind of Newton). *La Pensée* no. 259 (1987): 17–37.

————. *La Matière dérobée* (Matter revealed). Paris: Editions Archives Contemporaines, 1988.

————. "Matière, espace et temps selon Newton" (Matter, space, and time according to Newton). *Scientia* 107 (1972): 995–1026.

Pecker, J. C. *Clefs pour l'Astronomie* (Keys for astronomy). Paris: Seghers, 1981.

Piaget, J. *The Child's Conception of Time.* Trans. A. J. Pomerans. New York: Basic Books, 1969.

————. *The Construction of Reality in the Child.* New York; Basic Books, 1954.

Planck, M. *Autobiographie scientifique* (Scientific autobiography). Paris: Albin Michel, 1960.

Popper, K. R. *The Logic of Scientific Discovery.* New York: Harper & Row, 1968.

Prentzas, N., and M. Casse. "L'Avenir de l'univers" (The future of the universe). *La Recherche* 15, no. 156 (1984): 838.

Preskill, J. "Magnetic Monopoles." *Annual Review of Nuclear Science* 34 (1984): 461–530.

Prigogine, I. *From Being to Becoming: Time and Complexity in the Physical Sciences.* San Francisco: W.H. Freeman, 1980.

————. "The Rediscovery of Time." *Zygon* 19 (1984): 433–47.

Prigogine, I., and I. Stengers. *Entre le temps et l'éternité* (Between time and eternity). Paris: Fayard, 1988.

———. "La Nouvelle alliance" (The new alliance). *Scientia* 112 (1977): 287–304; *Scientia* 112 (1977): 617–30.

———. *Order out of Chaos: Man's New Dialogue with Nature.* New York: Bantam, 1984.

Prochiantz, A. *Les Stratégies de l'embryon* (The embryo's strategies). Paris: P.U.F. 1988.

Ramsey, N. F. "Electric Dipole Moments of Particles." *Annual Review of Nuclear Science* 32 (1982): 211–33.

Reinberg, A. *Des rythmes biologiques à la chronobiologie* (From biological rhythms to chronobiology). Paris: Gauthiers-Villars, 1977.

Reeves, H. *Atoms of Silence: An Exploration of Cosmic Evolution.* Trans. Ruth A. Lewis and John S. Lewis. Cambridge: MIT Press, 1985.

Reichenbach, H. *The Direction of Time.* Berkeley: University of California Press, 1971.

———. *The Philosophy of Space and Time.* Trans. Maria Reichenbach and John Freund. New York: Dover, 1958.

Reichenbach, H., and R. A. Mathers. "The Place of Time and Aging in the Natural Sciences and Scientific Philosophy." In J. E. Birren, ed. *Handbook of Aging and the Individual Psychological and Biological Aspects,* pp. 43–80. Chicago: University Chicago Press, 1959.

Richelle, M., and H. Lejeune. "L'Animal et le temps" (The animal and time). In Fraisse, ed., *Du temps biologique au temps psychologique,* p. 73.

———. "La Perception du temps chez l'animal" (The animal's perception of time). *La Recherche* 13, no. 182 (1986): 1332.

Ricoeur, P., ed. *Le Temps et les philosophies* (Time and philosophies). Paris: Payot/UNESCO, 1978.

Sachs, R. G. "Time Reversal." *Science* 176 (1972): 587–97.

Sadi Carnot et l'Essor de la thermodynamique (Sadi Carnot and the implications of thermodynamics). Paris: Editions du CNRS, 1976.

Salomon, M. *Future Life.* New York: Macmillan, 1983.

Schrödinger, E. *What Is Life? The Physical Aspect of the Living Cell and Mind and Matter.* New York: Cambridge University Press, 1967.

Segre, E. *From X-Rays to Quarks: Modern Physicists and Their Discoveries.* San Francisco: W.H. Freeman, 1980.

Shimony, A., and H. Feshbach, eds. *Physics as Natural Philosophy: Essays in Honor of Laslo Tizza on His Seventy-fifth Birthday.* Cambridge: MIT Press, 1981.

Strehler, B. L. *Time, Cells, and Aging.* New York: Academic Press, 1977.

Tank, D. W., and J. J. Hopfield. "Collective Computation in Neuronlike Circuits." *Scientific American* 257 (1987): 62–71.

Taton, R., ed. *Histoire générale des sciences* (General history of the sciences). Paris: P.U.F. 1958.

Teilhard de Chardin, P. *The Phenomenon of Man.* New York: Harper, 1975.

Tonnelat, M. A. *Histoire du principe de relativité* (History and the principle of relativity). Paris: Flammarion, 1971.

Thuan, T. X. "Le Big–bang aujourd'hui" (The Big Bang today). *La Recherche* 15, no. 151 (1984): 34.

———. "La Formation de l'univers" (The formation of the universe). *La Recherche* 17, no. 174 (1986): 172–80.

Thuan, T. X., and Th. Montmerle. "La Masse invisible de l'univers" (The universe's invisible mass). *La Recherche* 13, no. 139 (1982): 1438.

Varela, F. *Autonomie et connaissance: Essai sur le vivant* (Autonomy and conciousness: an essay on life). Paris: Seuil, 1989.

Wallis, R. *Time: Fourth Dimension of the Mind.* New York: Harcourt, Brace and World, 1968.

Wheeler, J. A. "Frontiers of Time." In *Problems in the Foundation of Physics,* ed. G. Toraldo di Francia, pp. 395–497. New York: North-Holland, 1979.

Whitrow, G. J. *The Natural Philosophy of Time.* New York: Clarendon, 1980.

Winfree, A. T. *The Geometry of Biological Time.* New York: Springer-Verlag, 1980.

Wolfenstein, L. "Present Status of CP Violation." *Annual Review of Nuclear Science* 36 (1986): 137–70.

Young, J. Z. *Programs of the Brain.* New York: Oxford University Press, 1978.

Zeman, Jiri, ed. *Time in Science and in Philosophy: An International Study of Some Current Problems.* New York: Elsevier, 1971.

Index

RÉMY LESTIENNE is research director at France's Centre National de la Recherche Scientifique (CNRS). An elementary particle physicist, he worked with the CERN accelerator in Geneva before turning his attention toward the neurosciences. Currently pursuing his investigation of temporal mechanisms in the brain at the Institut des Neurosciences in Paris, he is also the author of *Le Hasard Créateur* (Creative chance).

C. E. NEHER, who holds a B.A. in French from Coe College in Iowa and is the daughter of an American diplomat, was raised overseas and has studied in Paris. The coauthor of one published novel, she is currently working at the CNRS office at the Embassy of France in Washington, D.C.